Current Developments in Optical Fiber Technology

Current Developments in Optical Fiber Technology

Editor

Edvin Falk

scitus academics

Current Developments in Optical Fiber Technology
Edited by **Edvin Falk**

ISBN: 978-1-68117-287-3
Library of Congress Control Number: 2016937463

www.scitusacademics.com
Box No. 4766, 616 Corporate Way,
Suite 2, Valley Cottage,
NY 10989

Printed in United States of America on Acid Free Paper

Preface

Current optical fiber technology developments revolutionize world communication. Newly developed fiber amplifiers provide for the direct transmission of high-speed signals over transcontinental distances without electronic regeneration. Optical fibers find new applications for data processing. The impact of fiber materials, equipment and systems on communications in the coming decades will create a wealth of primary literature and the need for up to date reviews. An optical fiber is a flexible, transparent fiber made from glass (silica) or plastic that is slightly thicker than a human hair. A fiber optic cable is a bundle of glass threads that can be transmitted to light waves by modulated messages. Optical fibers are most often used to transmit light between the two ends of the fiber and are widely used in fiber optic communications where it is possible to transmit over longer distances and at higher bandwidths (data rates) than wire cables. Instead of metal wires, fibers are used because signals travel with fewer losses, and fibers are also immune to electromagnetic interference, a problem with which metal wires are excessively affected. Fibers are also used for illumination and wrapped in bundles for the transport of images, enabling viewing in confined spaces, as with a fiberscope. In a variety of other applications, some of them optical fiber sensors and lasers, specially designed fibers are also used.

This book is intended to offer contemporary developments and trends in optical fiber technology in connection with the next invention of optical communication, system and network, sensor, laser, measurement, characterization and equipment of fiber technology.

Table of Contents

CHAPTER 1

Advanced Optical Fibers for High Power Fiber Lasers

Liang Dong[1]

[1] Department of Electrical and Computer Engineering,
University of Clemson, Clemson, SC, USA

1. INTRODUCTION

The nonlinear limits of conventional single-mode fibers were well recognized in the nineties during the earlier efforts to scale the peak powers of pulsed fiber lasers. These earlier efforts focused on lowering the NA and optimizing the refractive index profile of optical fibers to increase the effective mode area [1, 2]. The emergence of photonic crystal fibers in the late nineties gave new impetus to the mode area scaling of single-mode optical fibers. The observation of the "endlessly single-mode" nature of *photonic crystal fibers* (PCF) at small hole sizes in 1996 [3] led to an early realization of the dispersive nature of a photonic crystal cladding [4], which limits the increase of normalized frequency V at short wavelengths. It was quickly realized that the scalability of Maxwell's equation allows for single-mode operation at a very short wavelengths in a small core to be directly translated into single-mode operation in a large core at longer wavelengths [5]. This led to a rapid progress in scaling of core diameters of single-mode PCF, culminating in the 100μm-core diameter demonstrated in 2006 [6].

The design and fabrication techniques developed for photonic crystal fibers also led to a realization of the potential of optical fibers with more open cladding structures than the closed concentric circles found in conventional optical fibers. It then became possible to design and fabricate leaky waveguides where differential mode losses can be used to control the number of propagation modes. As a consequence, *leakage channel fibers* (LCF) were proposed and first demonstrated in 2005 [7]. Two years later, single-mode operation in a record core diameter of ~180μm was demonstrated in a LCF [8].

Many other approaches based on conventional fibers were also pioneered during the last decade. One notable example is the *chirally coupled core* (CCC) fiber [9, 10], which relies on out-coupling of *higher-order modes* (HOM) to side cores adjacent to the main core. Phase-matching is achieved with the help of angular momentum from the helical side cores, which are formed by spinning the preform during fiber drawing. Another notable example is based on the propagation of a higher-order mode in a specially designed multimode fiber [11, 12]. It is argued that perturbations mostly have anti-symmetry in optical fibers and promote mode coupling mostly between modes of opposite symmetries. The mode spacing between a radially symmetric LP0n mode and its nearest neighbor modes with opposite symmetry is, in fact, larger for higher order modes. These higher-order modes are also more resistant to bend-induced mode compression. A special fiber design facilitates the ease of mode conversion to and from the LP0n mode using a *long period grating* (LPG).

In this chapter, we will give a brief introduction to the key approaches to effective mode area scaling which have shown great promise for future high power fiber lasers. Basic concepts are introduced and the latest developments are also discussed.

2. PHOTONIC CRYSTAL FIBERS

The authors of the paper in 1996 [3] were looking for photonic bandgap guidance in the defect core of the PCF with a core size of ~4.6 μm. The reason for the absence of photonic bandgap guidance in a solid-core PCF with air holes in the cladding only became clear a few years later. To their surprise, they observed very broadband single-mode guidance from 458–1550 nm (extended to 337-1550 nm in [4]). A conventional single-mode step-index optical fiber would become multimode at shorter wavelength. Even more surprising, the diffraction angle of the output beam was smaller at shorter wavelengths, indicating a smaller fiber numerical aperture at short wavelength.

2.1. Fundamental space-filling mode of photonic crystal cladding

In a follow-on paper [4] the initial observation of the "endlessly single-mode" nature of photonic crystal fibers was explained by the dispersive nature of the cladding, which can be viewed as a composite of two materials, i.e. silica and air. The effective refractive index of the composite cladding was identified for the first time as the effective index of the fundamental mode of the composite cladding, referred to as the *fundamental space-filling mode (FSM)*. In the extreme case of conventional optical fibers with an infinite cladding, the effective index of the fundamental space-filling mode becomes the refractive index of the cladding. In PCFs with air holes, the cladding index n_{cl}, i.e. the effective index of the fundamental space-filling mode, is lower than the refractive index of the background glass n_b due to the existence of the air holes, i.e. $n_{cl}<n_b$. The core index n_{co} is the same as the index of the background glass, $n_{co}=n_b$. Fundamental optical guidance in PCFs can, therefore, in principle, be understood in a similar way to that in conventional optical fibers (see Figure 1).

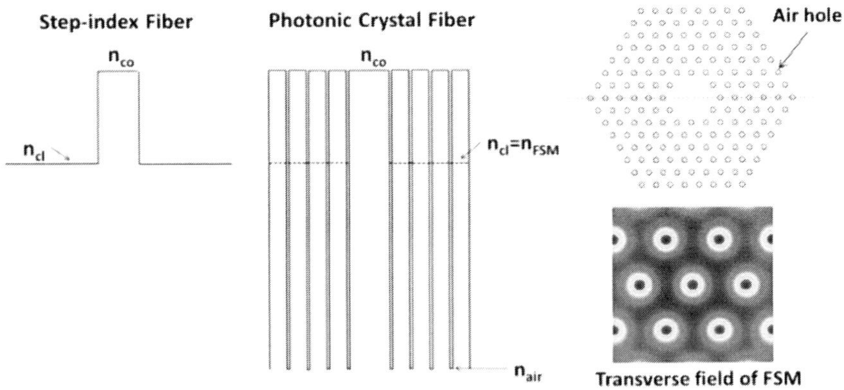

Figure 1. Refractive index of photonic crystal fiber cladding.

This paper [4] established the principle for understanding the basic guidance properties of PCFs. In conventional optical fibers, the refractive indexes of core and cladding are only weakly dependent on wavelength due to material dispersion. This is also expected of the core refractive index of a PCF. The refractive index of the composite photonic crystal cladding, however, behaves very differently. At the long wavelength extreme, i.e. $\lambda\rightarrow\infty$, the wavelength is much larger than the air holes and the fundamental space-filling mode will occupy all areas of the cladding equally. In this case, the refractive index of the FSM can be obtained by averaging the square of the refractive index of the composite photonic crystal cladding, i.e. $n_{cl}^2=n_b^2(1-F)+n_{air}^2F=n_b^2-F(n_b^2-n_{air}^2)$ where F is the area fraction of air; and n_{air} is the refractive index of air. The effective NA of the PCF for $\lambda\rightarrow\infty$ is, therefore, $F^{1/2}(n_b^2-n_{air}^2)^{1/2}$. At the short wavelength extreme, i.e. $\lambda\rightarrow0$, the wavelength is much smaller than the pitch of the air holes and the fundamental space-filling mode is mostly in the region with high refractive index, i.e. the background glass. In this case, the cladding index is the refractive index of the background glass n_b. The effective NA of a PCF at $\lambda\rightarrow0$ trends to zero. The guidance of a PCF still gets weaker at large λ, as in a conventional optical fiber, due to the inverse wavelength dependence of the normalized frequency V. In addition, its guidance also gets weaker at short wavelengths due to the diminishing NA, which also limits the growth of V and leads to the "endlessly single-mode" nature of the PCFs with small air holes. Guidance of PCF will, therefore, diminish at both long and short wavelength ends, possessing two bend-induced cut-off edges.

2.2. Single mode and multimode regime of photonic crystal fibers

In conventional fibers, a mode is guided when the effective mode index n_{eff} is between the core and cladding index, i.e. $n_{co}>n_{eff}>n_{cl}$. The mode cut-off can be identified when $n_{eff}=n_{cl}$. The second-order mode cuts off in a step-index fiber at V=2.405. In the first reported work on determining the single-mode regime of PCFs with a 1-cell core, i.e. one hole missing at the defect core, by Birks et al [4], a somewhat arbitrary equivalent step-index core radius equaling *pitch*Λ, i.e. center-to-center hole separation, was used to calculate the V value. Work by Saitoh [13] using a *finite element model* (FEM) to determine n_{FSM} and the condition $n_{eff}=n_{cl}$ to determine the second-order mode cut-off, put the effective step-index core radius to be $\Lambda/3^{1/2}$ for a one-cell core. Later work by Saitoh [14] determined the effective step-index core radius to be Λ for a 3-cell core and $2^{1/2}\Lambda$ for a 7-cell core.

The optical waveguide equation allows the scaling of all parameters measured in the length scale by the same factor. For PCF, the most convenient scaling factor is $1/\Lambda$. The second-order mode cut-off is typically plotted as a normalized wavelength λ/Λ versus normalized hole diameter d/Λ plot. This is shown for PCFs with 1-cell, 3-cell and 7-cell core in Figure 2 [14]. For each of the curves, the area above the curve is in the single-mode regime, i.e. the wavelengths above the second-order mode cut-off wavelength, and, below it, multimode regime. The "endlessly single-mode" nature of PCFs can be easily identified in Figure 2. When $d/\Lambda<0.424$, 0.165 and 0.046 respectively for 1-cell, 3-cell and 7-cell core PCFs, the PCFs will remain in the single-mode regime for all wavelengths. It is worth noting that Figure 2 is for PCFs with infinite cladding. For PCFs with finite cladding, guidance is weaker and the second-order-mode cut-off is expected to happen at slightly shorter wavelengths. The curves in Figure 2 are expected to move downwards slightly. It needs to be noted as discussed earlier, that the PCFs are too weak to guide any light at long and short extremes of wavelength. The critical bend radius at the short wavelength edge is determined to be dependent on pitch Λ and wavelength λ such that [4]

$$R_c \propto \frac{\Lambda^3}{\lambda^2} \tag{1}$$

The critical bend radius at the short length edge increases in proportion to $1/\lambda^2$ as wavelength decreases. This relation was verified experimentally for the critical bend radius at 3dB excessive bend loss [4]. It comes directly from the dispersive nature of the photonic crystal cladding.

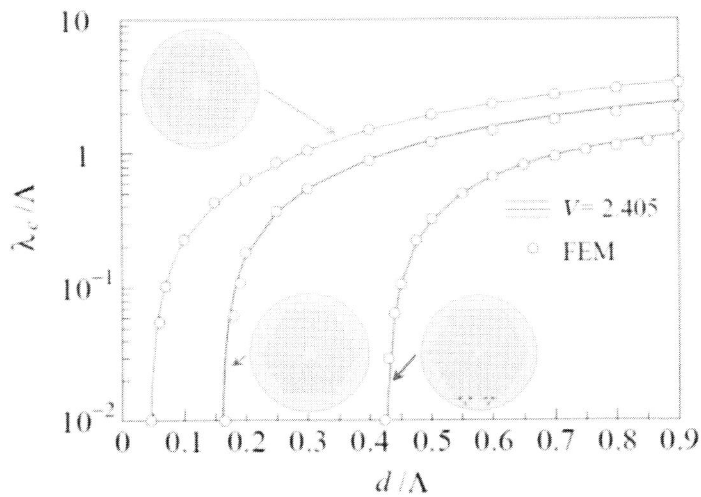

Figure 2. The second-order mode cut-off determined by $n_{eff}=n_{FSM}$ (dots) and V=2.405 (solid lines) using core radius $\rho=\Lambda/3^{1/2}$, Λ and $2^{1/2}\Lambda$ respectively for 1-cell, 3-cell and 7-cell cores [14].

2.3. Waveguide loss of photonic crystal fibers

PCFs are intrinsically leaky, i.e. there is always a finite waveguide loss for each mode in a PCF with a finite cladding. The waveguide loss can be found by calculating the imaginary part of the effective mode index using a numerical mode solver. When plotting the waveguide loss versus wavelength, the slope is expected to change around cut-off. This can also be used to determine the second-order-mode cut-off. This was done for a 1-cell core PCF [15]. The results are consistent with those in [13].

In conventional optical fibers, the fully enclosed core and cladding boundary ensures that waveguide loss is zero for all guided modes, i.e. those that satisfy the conditions for total internal reflection at the boundary. For the open structure of PCFs with a finite cladding, all modes are leaky with finite waveguide losses. At the extreme of an infinite number of layers, the waveguide losses are zero. The waveguide losses are lower for larger air holes and can be substantially lowered by increasing the number of air-hole layers. In practice, the waveguide loss can be made below other material and process related losses. By employing appropriate polishing, etching and dehydration processes, a PCF with a loss 0.28dB/km at 1550nm has been demonstrated [16]. For applications in fiber lasers where a length of not more than a few tens of meters is used, the waveguide loss does not present any problem if appropriate designs are used.

2.4. Modeling of photonic crystal fibers

Due to the complexity of the geometrical structures of PCFs, numerical models are typically used to find mode properties including the effective mode index, waveguide loss and effective mode area. For any numerical mode solvers, electric and magnetic fields are described by an expansion series. Eigenvalue equations are then established by requiring the fields to satisfy all boundary conditions. These equations are typically expressed as a set of linear equations, which can be solved for complex effective mode index. The more complex a waveguide design is, the larger is the number of linear equations. The waveguide dispersion can be calculated from the real part of the complex effective mode index. The waveguide loss can be obtained from the imaginary part of the complex effective mode index. Electric and magnetic fields can be calculated once the effective mode index is determined. The most common and flexible numerical mode solver is a *finite element mode solver* (FEM). This is commercially available from COMSOL (*http://www.comsol.com/*). A FEM is capable of handling practically any design. It can, however, be computationally very demanding. If only circular boundaries are involved, a Multipole mode solver is a good option [17, 18]. It is based on the decomposition of fields into circular Bessel series, which are the most accurate and efficient method for modeling circular boundaries. For non-commercial research and teaching purposes, it can be downloaded from the University of Sydney website (*http://sydney.edu.au/science/physics/cudos/research/mofsoftware.shtml*). A plane wave expansion method can also be used. This, however, assumes an infinite cladding and, therefore, cannot determine the waveguide loss.

2.5. Mode area scaling with photonic crystal fibers

The first demonstration of a large-core PCF was performed by Knight el at [5]. The 1-cell PCF with relative hole diameter $d/\Lambda \approx 0.12$ and a core diameter of $2\rho = 22.5\mu m$, provided robust single-mode guidance at 458nm. According to Figure 2, a 1-cell PCF with $d/\Lambda < 0.424$ is expected to be single-mode over its entire wavelength range. It is, therefore not surprising that the PCF was single mode at 458nm. The critical bend radius was measured to be 50cm at 458nm and 4cm at 1.55μm [5]. This fiber is expected to have a critical bend radius of ~10cm at 1μm using equation 1.

Considering the wavelength scalability of the waveguide equation, this 1-cell PCF with $2\rho/\lambda \approx 50$ can be scaled by a factor of ~2.2 to a ~50μm core diameter to operate at 1μm. The critical bend radius for this single-mode PCF with 50μm core diameter is expected to be ~1.1m at 1μm according to equation 1! A similarly scaled 1-cell PCF with a core diameter of 30μm will have a critical bending radius of ~24cm. This is probably close to the practical limit of coiled single-mode 1-cell PCFs. Above a core diameter of ~40μm, 1-cell single-mode PCF can only be used in a straight configuration in practice. The single-mode operation of PCFs with large core diameters comes at the cost of weak guidance as a result of the diminishing effective NA. This fundamentally limits the use of single-mode PCFs with large cores in coiled configurations. If PCFs are allowed to operate in the few-moded regime, coiled PCFs with slightly larger core diameters are possible. For high average power fiber lasers with outputs exceeding kW, effective thermal management becomes increasingly critical. Long fiber lengths of many meters are required. The constraint of not being able to coil the fibers can become a major issue considering the additional space constraints.

2.6. Rare-earth-doped glass for large-core photonic crystal fibers

The diminishing NA of PCFs with large cores also has additional implications, as realized by the authors of [5]. To fabricate rare-earth-doped large-core PCFs, the core refractive index needs to be accurately controlled at levels far below the very small effective index difference between the core and cladding in order not to disturb the guidance properties of the PCFs. This requires much improved techniques for the fabrication of the active core in large-core PCFs.

The first ytterbium-doped large-core PCF with 15μm core diameter was reported in 2001 at Bath University [19]. The effective mode area at 1μm was ~100μm^2. The laser operated in single mode with poor efficiency. The key advance was the use of a repeated stack-and-draw process to achieve a uniformly doped core with a refractive index close to silica. Conventional fabrication techniques for rare-earth doped silica fibers result in significant non-uniformity in refractive index across the core as well as a raised refractive index above silica. Due to the weak guidance in large-core PCFs, any index non-uniformity across the core can lead to mode distortion. It also requires the refractive index of the core to be very close to the silica background in PCFs. The authors of [19] fabricated ytterbium-doped glass with low doping levels. The doped glass is surrounded by some undoped silica glass. The glass is stacked and drawn twice before being finally incorporated into the core of a PCF. The resulting PCF has an ytterbium-doped core which consists of 425 original doped glass sections. The dimension of each of the original doped glass sections is much smaller than the wavelength of light and the core of the PCF, therefore, appears to be homogeneous to light at the operating optical wavelength. The doped core is also heavily diluted by the addition of silica glass (90%), leading to an average refractive index close to that of silica [20]. In the same paper, it was pointed out that rare-earth-doped glass with a higher index can be stacked and drawn with undoped glass with a low index to achieve a better match to that of silica background [20].

2.7. Double-clad photonic crystal fibers with high NA air-clad for pump guidance

The concept of a composite air-glass clad with a high air-filling fraction to provide a high NA pump guide in a double-clad fiber was first proposed in 1999 at what was then then Lucent Technologies [21]. The high pump NA can enable a significant improvement in pump coupling especially from low brightness multimode diode lasers for a given pump waveguide dimension. In conventional double-clad fibers, low-index polymer coatings are typically used to form the pump cladding. The air-clad pump isolates high pump powers from the polymer coatings, leading to potentially improved reliability.

The first such fiber was demonstrated in 2000, where the pump cladding was mostly air except for a single connecting element [22]. The measured pump NA over 1m was below 0.2. The fiber was passive and there was no laser demonstration in this first attempt. The first cladding-pumping demonstration in an active fiber with an air-clad pump guide and a conventional core design was in 2001 at Southampton University [23]. The measured NA of the pump guide with air-cladding was 0.4-0.5. In a separate demonstration from Southampton University later in the same year, a conventional core with a very low NA together with a photonic crystal cladding was used in combination with a pump guide with air-cladding [24]. The core guidance came from a combination of the raised core index and the photonic crystal cladding. Cladding pumping with a low brightness laser diode at 915nm (100μm core with a NA=0.22) was demonstrated with a slope efficiency of 70% relative to the absorbed pump power. The measured pump NA over a short length (~10cm) was 0.3-0.4. NA over much longer fiber length as in a fiber laser is expected to be lower. The first true active double-clad photonic crystal fibers with a pump air-cladding was demonstrated in 2003 with a measured pump NA of 0.8 at 1μm [20].

The theoretical basis for the high NA of optical waveguides formed by air-cladding was established in [25]. It had been understood previously that the effective index of a glass and air composite can be obtained by the effective index of the fundamental space filling mode. An example of high NA air-clad used in double-clad fibers is shown in Figure 3 from [25]. In such air-glass composites, the glass webs can be considered as slab waveguides. The separations between the webs are typically far enough that the webs can be considered as isolated. In this case, the effective index of the glass-air composite cladding can be well approximated by that of the fundamental slab waveguide mode. This can be easily calculated. The resulting pump NA is only dependent on the ratio of the web width w and wavelengthλ. This is plotted in Figure 3(c), showing that a small w/λ is critical for high NA.

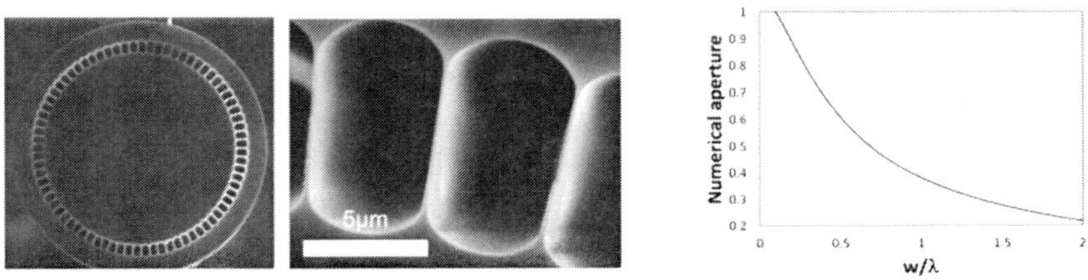

Figure 3. (a) High NA air-clad from [25], (b) close-up of the high NA air-clad, and (c) NA of air-clad waveguide based on slab waveguide model plotted against the ratio of waveguide width w and wavelength λ.

2.8. Progress of active large-core photonic crystal fibers

Since 2003, the group at Friedrich-Schiller-University Jena has been playing a significant role in the development large-mode-area photonic crystal fibers with pump air-clad. In their first work in collaboration with then Crystal Fibre A/S, now NKT Photonics, an ytterbium-doped PCF with effective mode area of ~350μm^2 and mode field diameter (MFD) of 21μm was demonstrated [26] (see Figure 4). The 3-cell PCF had a hole diameter d=2μm and pitch Λ=11.5, giving an d/Λ=0.18. The core diameter was 28μm. A relatively small circular area of 9μm was doped in the 3-cell core. The doped area had a high ytterbium-doping level of ~0.6 at%. It was further co-doped with aluminum and fluorine to provide an index merely 2×10^{-4} above the silica background. The PCF with the raised index in the doped area was simulated, showing that the fiber guides the second-order mode, which is, however, close to cut-off. The fiber had a 150μm pump guide with an air clad. The webs in the air clad were ~50μm long with a thickness of ~390nm, giving w/λ=~0.4 at 976nm. The measured pump NA was 0.55, slightly below the NA=0.68 predicated by the slab model in Figure 3. The fiber had an outer glass diameter of 450μm and was coated with standard acrylic coating. Due to the high doping level, the fiber has a pump absorption of ~9.6dB/m at

976nm. An impressive slope efficiency of 78% was demonstrated with respect to the launched pump power.

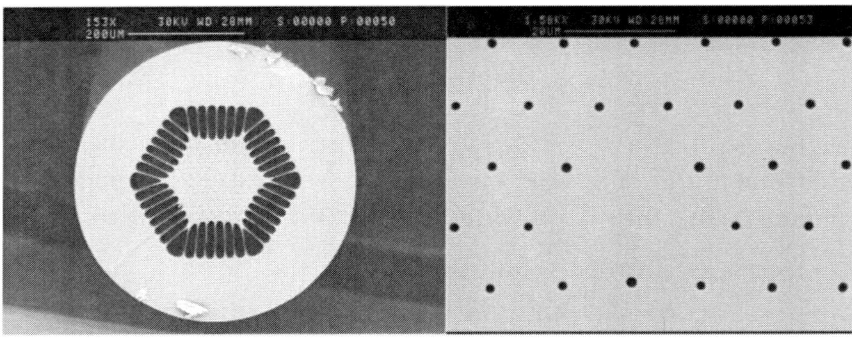

Figure 4. SEM image of the air-clad ytterbium-doped large-mode-area photonic crystal fiber in [26].

In a subsequent paper in 2003 [27], the fiber in [26] and a new fiber with narrower pump air-clad were studied with the FEM for temperature distributions in the fiber at various thermal loads in the core, considering both convective and radiative heat transportation by air at the fiber surface. The results show that, in the case of natural cooling, the thermal transportation is mainly limited by the heat transfer at the fiber surface. The air-clad impedes heat flow, especially when the width of the air clad is large. A narrower pump air-clad is advantageous, especially in actively cooled fiber lasers.

In 2004, the bar was raised in a collaborative work between Friedrich Schiller University Jena and Crystal Fibre A/S [28]. Their 7-cell fiber has an effective area of ~1000μm^2 and MFD of ~35μm (see Figure 5). Core diameter is ~40μm. Hole diameter d is 1.1μm and the pitch Λ is 12.3μm, giving a d/Λ=0.09. The fiber has a pump guide diameter of 170μm and a measured pump NA of 0.62 at 950nm. The pump absorption is ~13dB/m at 976nm. The pump air-clad is much narrower for more efficient heat diffusion. It is worth noting that the fiber is in the multimode regime for a 7-cell PCF (see Figure 5). This helps to ease bend loss in the weakly guided fiber.

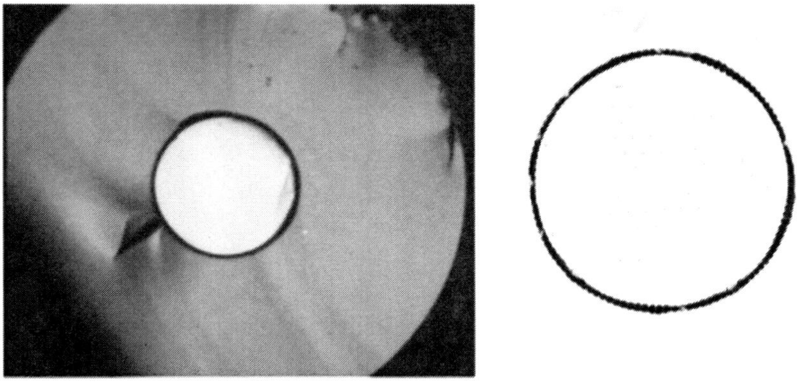

Figure 5. Microscope image of the air-clad ytterbium-doped single-mode PCF and close-up to the 40-μm core formed by seven missing air holes [28].

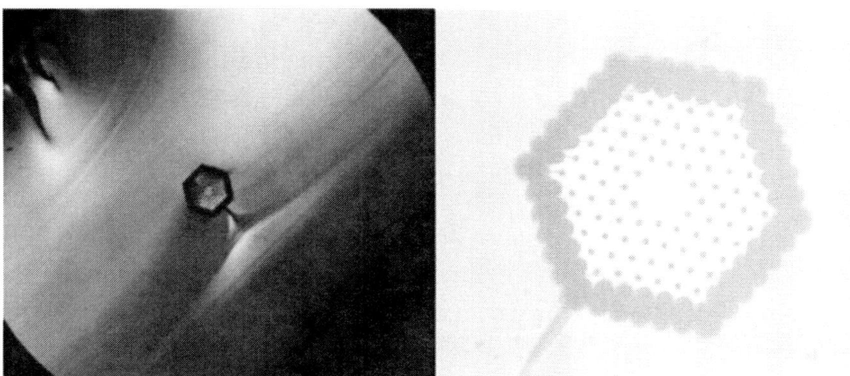

Figure 6. Microscope image of a rod-type photonic crystal fiber and close-up view of the inner cladding and core regions.

To further mitigate bend loss, a rod-type PCF was developed in 2005 [29] (see Figure 6). The rod had an outer diameter of 1.7mm so that it cannot be bent. A 48cm long length was used in the demonstration. The relative hole diameter was increased to $d/\Lambda=0.33$. This reduces bend sensitivity. This larger d/Λ in a 7-cell core PCF, however, puts this fiber firmly in the multimode regime (see Figure 2). The pump guide was reduced to a hexagon with flat-to-flat dimension of 117μm and corner-to-corner dimension of 141μm, similar to earlier fibers. The ytterbium doping level was also increased compared with earlier fibers. The increased doping level and reduced pump guide led to a high pump absorption of ~30dB/m at 976nm. The fiber was used to demonstrate power extraction of ~250W/m.

The rod-type PCF was further developed with the demonstration of a 19-cell PCF with a core diameter of 60μm, effective mode area of ~2000μm^2 and MFD of 75μm in 2006 [30] (see Figure 7). The fiber had a $d/\Lambda=0.19$, again firmly in the multimode regime (19-cell PCF not shown in Figure 2). The pump guide was 175μm in diameter. The pump air-clad had a web thickness of 400nm, which was 10μm long. The measured pump NA was 0.6 at 975nm, fairly close to the 0.67 NA predicted by the slab model (see Figure 3). The pump absorption is ~30dB/m at 976nm. The rod diameter is ~1.5mm. A 58cm long fiber was used to demonstrate 55W/m power extraction. In the same paper, a passive 100μm core fiber was also demonstrated. The 19-cell fiber had a $d/\Lambda=0.2$, again in the multimode regime. The effective mode area was ~4500μm^2 and MFD was ~75μm.

The demonstration of a 100μm active 19-cell PCF finally came in 2006 [6]. The pump guide had a diameter of 290μm. The rod diameter was 1.5mm. A 90cm long fiber was used to amplify 1ns pulses at 9.6kHz to record peak power of 4.5MW and pulse energy of 4.3mJ with $M^2=1.3$.

Polarization-maintaining (PM) PCFs have also been developed by introducing stress elements. A PM 7-cell PCF with a mode area of ~700μm^2 was demonstrated in 2005 [31] (see left figure in Figure 8). The pitch Λ was 12.3μm with a d/Λ of 0.2. In the weakly guided PCFs, the polarization modes on slow and fast axis have different bend losses.

This enables single-polarization operation where the polarization mode on the slow axis is still guided while polarization mode on the fast axis suffers high bend loss. Another single-polarization 19-cell PCF was demonstrated in 2008 with a mode area of ~2300μm^2 (right figure in Figure 8) [32]. The corner-to-corner distance of the core was 70μm. The pitch Λ was 11μm with a d/Λ of 0.1.

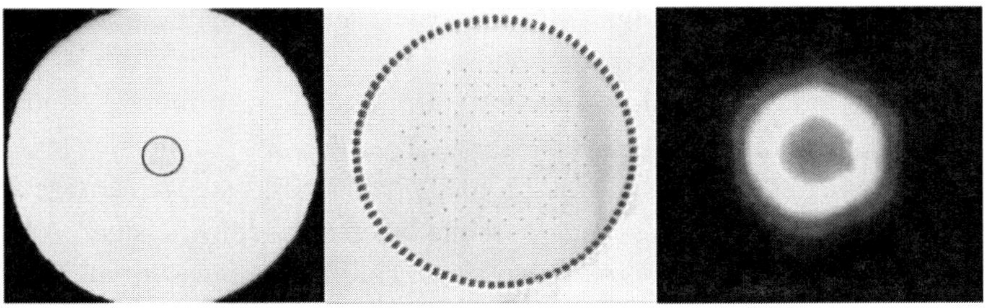

Figure 7. Microscope-image of the extended-mode-area rod-type photonic crystal fiber, SEM-picture of the micro-structured region and measured near-field intensity profile of the 60μm core fiber [30].

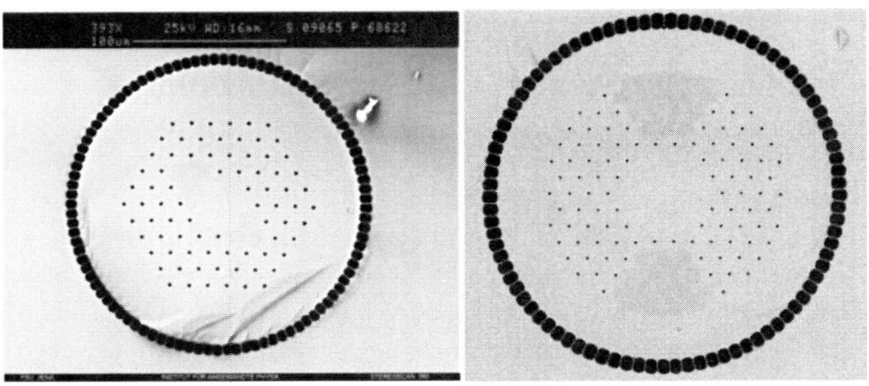

Figure 8. SEM image of the single polarization PCF with an effective area of ~700μm^2 [31] (left) and 2300μm^2 (right) [32].

3. LEAKAGE CHANNEL FIBERS

A 2D micro-structured cladding, which is made possible by the stack-and-draw technique developed for photonic crystal fibers, enables new designs which do not possess the closed core-and-clad boundaries of conventional optical fibers. When a mode is guided in a conventional optical fiber, total internal reflection everywhere around the closed core-and-clad boundary, traps light entirely in the core, leading to zero waveguide loss. In designs with an open cladding, light can leak out, leading to finite waveguide loss associated with each mode. The waveguide loss is mode-dependent, providing opportunities for mode control by minimizing loss of the desired mode while maximizing loss of the unwanted modes. *Leakage channel fibers* (LCF) takes advantages of these new opportunities made possible by open cladding designs. A LCF can be precisely engineered to have high confinement loss for all higher order modes and low confinement loss for the fundamental mode and can, therefore, significantly extend the effective mode area of the fundamental mode. LCFs essentially exploit the increased ability of higher order modes to leak through small gaps in the cladding while maintaining good fundamental mode confinement.

3.1. Leakage channel fibers with air holes

The first LCF was demonstrated in 2005 [7]. It has a simple cladding design with air holes in the cladding [7]. The LCF is shown in in the left figure in Figure 9. The LCF had an outer diameter of ~270μm. The two smaller holes had a diameter of d=39.6μm and pitch Λ=51.2μm. The four larger holes had a diameter of d=46.0μm and pitch Λ=51.1μm. The passive LCF provided robust single-mode operation with a measured mode area of ~1417μm^2 (MFD=42.5μm). The most significant aspect of this work is that the LCF can be coiled down to 15cm with negligible loss penalty. This is a significant improvement over PCFs.

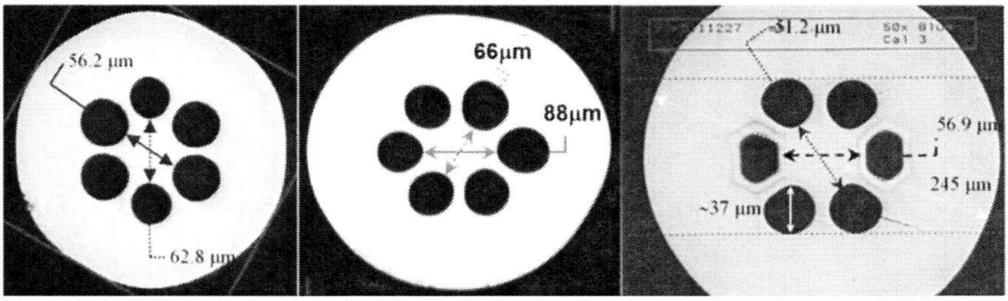

Figure 9. The LCF used in the first demonstration [7] (left), the first ytterbium-doped LCF (center) [33] and the first ytterbium-doped PM LCF (right) [34].

The first Ytterbium-doped LCF was demonstrated in 2006 [33] (see the middle figure in Figure 9). The fiber outer diameter was ~350µm and it was coated with a low index polymer to give a pump NA of 0.46. The average hole diameter was ~55µm and average pitch was ~67µm, giving an average $d/\Lambda=0.82$. The effective mode area was 3160µm^2 (MFD=63.4µm). Pump absorption was measured to be ~3.6dB/m at 976nm. Slope efficiency versus launched power was measured to be ~60% in a 5m long amplifier coiled at 40cm diameter. M^2 was measured to be ~1.3.

The first ytterbium-doped PM LCF was also demonstrated in 2006 [34] (see the right figure in Figure 9). A pair of boron-doped silica stress rods was used to replace two opposing air holes. The hole diameter was ~37µm. The effective mode area was ~1400µm^2 (MFD=42.2µm). The birefringence was measured to be $\sim 2.1\times10^{-4}$ over 1010-1080nm. The LCF had an outer diameter of ~245µm and was coated with a low index polymer to give a pump NA of ~0.46. The pump absorption was ~2.6dB/m at 976nm. Slope efficiency versus launched power was measured to be ~60%. M^2 was measured to be ~1.2. The PM LCF could be coiled to 12cm diameter with negligible bend loss.

3.2. All-glass leakage channel fibers

High refractive index contrast is not necessary for large core fiber designs. Low refractive index contrast is sufficient and often advantageous for further limiting higher order mode propagation. Fluorine-doped silica can be used to replace air holes in the LCFs described in the last section. The all-glass LCFs can provide much improved ease of fabrication and use, compared with fibers with air holes.

Despite the fact that air is a readily available ingredient, there are a number of drawbacks related to the use of air holes in fibers. The first one is the difficulty in precisely controlling the dimension of air holes in fiber fabrication. This is an intrinsic problem of a holey structure due to the air hole's tendency to collapse during fiber drawing. This is usually countered by a precise control of pressurization of the air holes, a process dependent on drawing conditions such as furnace temperature, feed rate, and drawing speed. When small air holes are desirable as in endless single-mode PCFs, higher pressure is required to maintain air hole dimensions due to the significantly increased tendency for the air holes to collapse by surface tension in this regime. This can make air hole dimensions to become highly sensitive to drawing conditions. This delicate balance of pressurization and collapse can lead to issues of controllability and repeatability in PCF fabrication. Air holes can also disturb smooth fracture wave propagation during fiber cleaving, leading to a poor cleaved surface due to the appearance of deep fractures behind the air holes, a problem often aggravated by large air holes and high cleaving tensions. In addition, air holes often have to be thermally sealed at the fiber ends to minimize environmental contamination. Mode distortion can occur from the air holes collapsing during splicing. This is especially true for large-mode-area fibers, which are much more susceptible to small perturbations.

A detailed analysis of all-glass LCFs was reported in [35]. For a LCF formed by one layer of features as shown in the inset of Figure 10, the core of diameter 2ρ is formed by six features with diameter d and refractive index n_f. Center-to-center feature spacing is Λ. The refractive index of the background glass, usually silica, is n_b. The fiber was studied by a multipole mode solver for the effect of index contrast $\Delta n = n_b - n_f$. For the simulation in Figure 10, the following parameters were used: $2\rho=50$ μm, $d/\Lambda=0.675$, and $n_b=1.444$. The wavelength of the simulations was at 1.05μm. It can be seen from the left figure in Figure 10 that confinement loss for both fundamental, α_{FM}, and second-order modes, α_{2nd}, increases with a reduction of index contrast Δn, with the loss of the second-order mode, α_{2nd}, remaining over an order of magnitude higher than the loss of the fundamental mode, α_{FM}. The modal index difference, the difference between the effective mode indices of the fundamental and second modes, decreases toward low index with Δn by just ~40% over three orders of magnitude change in Δn.

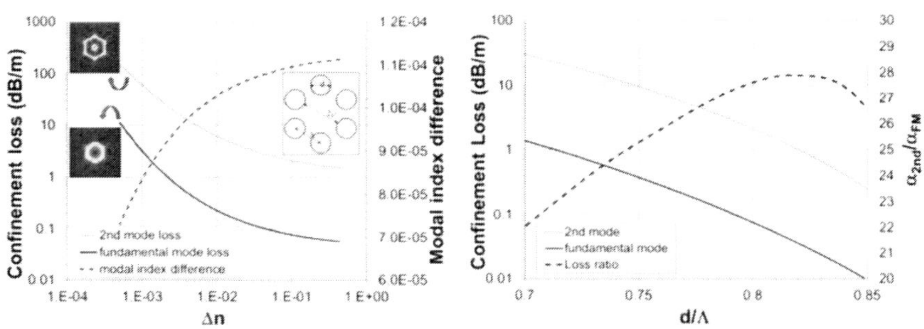

Figure 10. Effect of index contrast on confinement loss and modal index difference for one-cladding-layer LCFs with $d/\Lambda=0.675$ and $2\rho=50$μm (left) and effect of d/Λ on confinement loss and the loss ratio between the second-order mode and fundamental mode for LCFs with $\Delta n=1.2\times10^{-3}$ and $2\rho=50$μm (right) [35].

The effect of normalized hole diameter d/Λ was also studied in [35] and is shown in the right figure in Figure 10 for confinement losses and the ratio of the second-mode loss to the fundamental-mode loss. The confinement loss for both the fundamental and second modes increases toward small d/Λ with the loss ratio changing very little over the entire range of d/Λ, from 22 to 28. The normalized hole diameter d/Λ is typically chosen to give an acceptable fundamental-mode loss. For LCFs with one layer of features as shown in Figure 10, the loss ratio of all-glass LCFs is very similar to that of an LCF with air holes. Slightly larger d/Λ is, however, required for achieving a similar confinement loss.

LCFs with two layers of features can be used to further improve the differential confinement loss between the fundamental and second-order modes at the expense of bending performance. Acceptable fundamental-mode loss at smaller feature sizes can be realized in LCFs with two layers of features, while leakage of higher order modes is substantially increased by a reduction of feature size despite the additional layer of features. Higher differential loss between modes can therefore be realized. Since bending loss of the fundamental mode is very strongly dependent on feature size, a reduction of feature size increases the bend loss of the fundamental mode in LCFs. An LCF with two layers of features was studied in [35] and the results are shown in Figure 11. Both the fundamental mode and second-order-mode loss shows the characteristic increase at small d/Λ, while the loss ratio α_{2nd}/α_{FM} is increased by over an order of magnitude compared to the one-layer designs in Figure 10. At $d/\Lambda \approx 0.548$, the fundamental-mode loss $\alpha_{FM} \approx 0.1$ dB/m, while the second-order-mode loss $\alpha_{2nd} \approx 48$ dB/m, a loss ratio α_{2nd}/α_{FM} of ~480. A very high loss ratio α_{2nd}/α_{FM} of ~700 is possible at $d/\Lambda=0.62$.

A wide range of all-glass LCFs were fabricated from core diameters from 35μm to well over 100μm [35] (see Figure 12). All fibers were made with silica glass as the background glass and slightly fluorine-doped silica glass as the cladding features and coated with standard coating with index of 1.54. The refractive index difference between the background and the low index feature wasΔ $n=1.2\times10^{-3}$. LCFs with both circular and hexagonal features were fabricated and tested. The conditions for the fabrication of LCFs with hexagonal features also created LCFs with a rounded hexagonal outline. Such a shape is known to be preferred for the pump mode mixing in a double clad fiber where the pump light propagates in a much larger cladding guide. All the fabricated LCFs in Figure 12 operated in the fundamental mode with a varying degree of bend loss performance. In general, bend loss increases rapidly with core diameter increase (see Figure 13). This effect is fundamentally related to the fact that the ability of guided modes to navigate a bend is related to how rapidly a mode can change its spatial pattern without breaking up while propagating, i.e. maintain adiabatic transition. As the mode gets larger, this ability to change diminishes very quickly due to larger Rayleigh range.

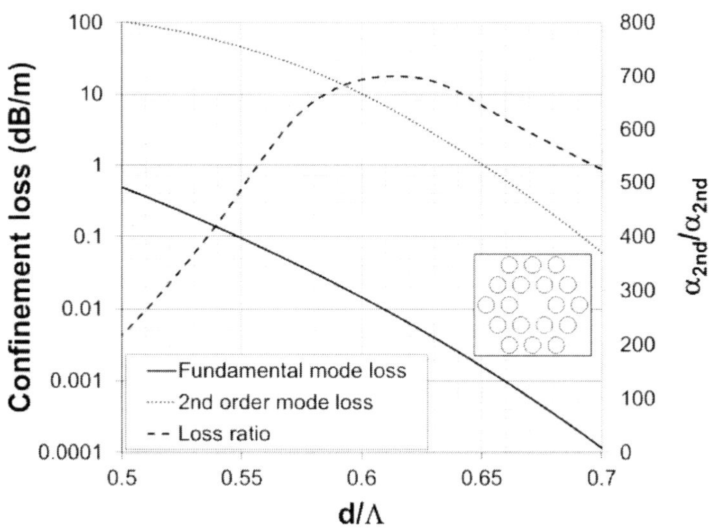

Figure 11. Effect of d/Λ on confinement loss and the loss ratio between the second-order mode and fundamental mode for an LCF with two layers of features, $\Delta n=1.2\times10^{-3}$ and $2\rho=50\mu m$. [35].

The LCF, shown on the top left inset in Figure 13, had $2\rho=101\mu m$, and $d/\Lambda=0.9$. The effective mode area of the LCF was calculated to be $5117\mu m^2$ (MFD=80.7μm). A length of the LCF ~6m long was loosely coiled in a 1m coil and the measured M^2 was $M^2_x=1.26$ and $M^2_y=1.29$. The LCF, shown on the top right inset, had $2\rho=183.3\mu m$, and $d/\Lambda=0.8$. A conventional single mode optical fiber of the same scale is also shown for comparison. The effective mode area of this LCF was calculated to be 15861μm^2 (MFD=142.1μm), a record effective mode area for single-mode operation. The Measured M^2 of a 1m straight fiber is M^2_x=1.22 and M^2_y=1.23. Measured mode patterns at the output of the fibers are also shown in Figure 13.

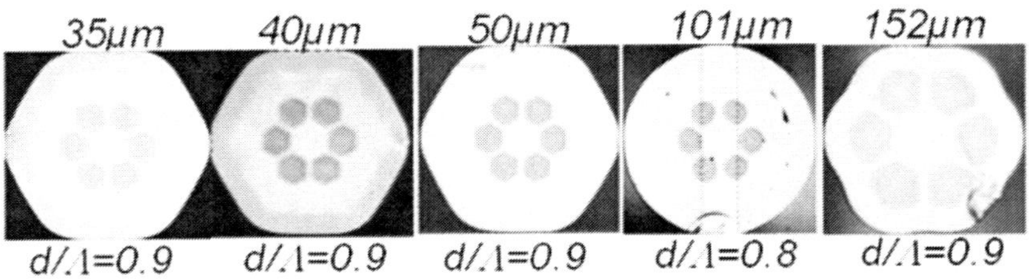

Figure 12. Some examples of fabricated all Glass LCFs. Core diameter is given above the fiber [35].

3.3. Polarization maintaining all-glass leakage channel fibers

A PM all-glass LCF was first reported in [37] (see Figure 14). The passive all-glass PM LCF had a core diameter of 50μm. A high d/Λ=0.9 was used for a smaller critical bend radius. The LCF had a refractive index difference between the background and the low index feature of $\Delta n=1.2\times10^{-3}$. The low index features were made of slightly fluorine-doped silica. Two stress elements with a refractive index of $\sim 13 \times 10^{-3}$ below that of the background silica glass were used instead of the regular features on either sides of the core to provide birefringence. The fiber had an outer diameter of ~885μm and was coated with standard acrylic coating. The near field image measured with a single lens is shown in Figure 14(c) for the 1.8m long sample and in Figure 14(d) for the 30m sample. Due to the much higher d/Λ=0.9 used for this fiber, some bending was necessary for fundamental mode operation in a short length of this fiber. The output was robustly single mode in a 30m long sample of this passive LCF coiled in 40cm diameter coils (see Figure 14(d)). The critical bend radius for 1dB/m loss, expected to be ~11cm by FEM simulation, matched very well to the 10.5cm measured. *Polarization extinction ratio* (PER) was characterized at the output of a 1.8m long sample to be >15dB over 1010-1100nm.

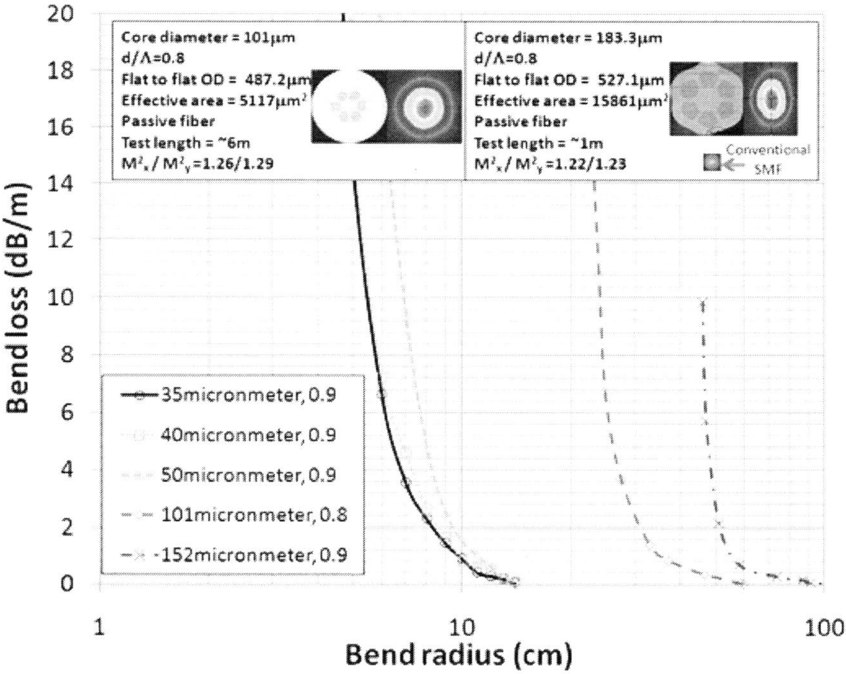

Figure 13. Cross section, measured mode and fiber details are given for the LCF with 101μm core, left inset, and the LCF with 183.3μm core, right inset. Measured bend loss for LCFs with various core diameters [36].

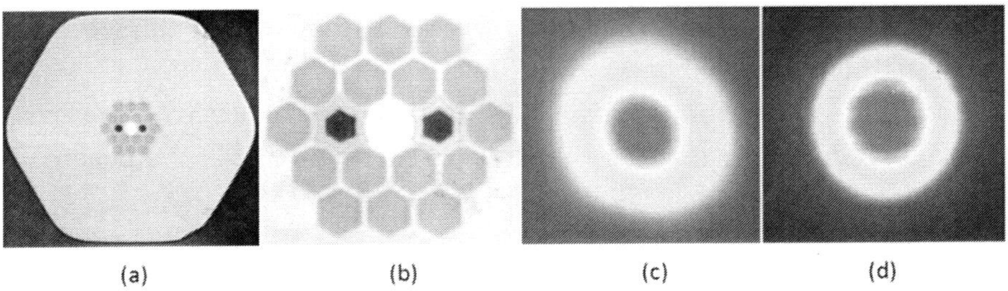

Figure 14. (a) Cross section of the passive PM LCF, (b) magnified cross section, (c) near field from the 1.8m long fiber in a 40cm-diameter coil, (d) near field from the 30m long fiber in a 40cm-diameter coil [37].

3.4. Characterization of mode losses in all-glass leaky channel fibers

Recently, fundamental and higher order mode losses have been characterized and compared to simulations based on the assumption of an infinite cladding [38]. A LCF with a ~50μm core diameter and a hexagonal cladding boundary (see Figure 15(a)) was spliced to a tunable source to ensure launch stability during the measurements. The passive LCF was coated with a low-index polymer to simulate a double-clad fiber with a pump NA of ~0.45. Power in various modes at the output was measured using the S^2 method [39]. The fiber was cut back several times to determine the propagation losses of various modes. The measurement was repeated at various coil diameters. The results are summarized in Figure 15(b) and show remarkable agreement between the measured and simulated losses of fundamental and higher-order modes. LP11 mode loss as high as ~20dB/m was measured, demonstrating the validity of the design. A similar LCF with a circular cladding boundary was also measured, showing significant less higher-order mode losses than those predicted by simulations. It is speculated that the coherent reflection at the circular cladding boundary played a significant role in enhancing the guidance of leaky higher-order modes in this case. It is, therefore, critical to have a non-circular cladding boundary to achieve the maximum possible higher-order mode losses.

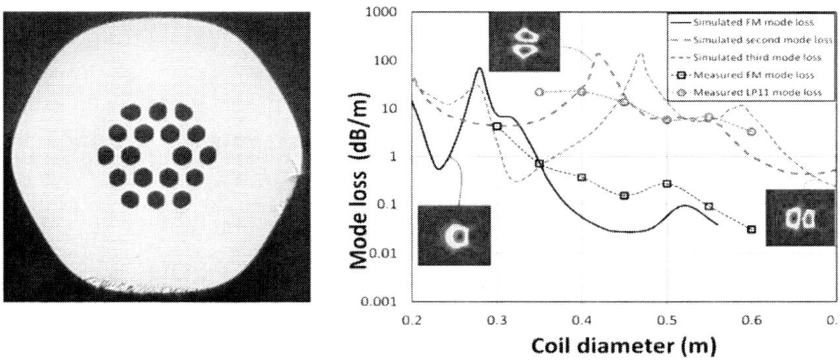

(a) (b)

Figure 15. (a) The hexagonal LCF used in the measurement and (b) simulated and measured mode loss in the hexagonal Re-LCF [38].

3.5. Ytterbium-doped all-glass leakage channel fibers

An ytterbium-doped all-glass LCF with one layer of cladding features was also reported in [37]. The LCF was coated with a low index polymer, providing a pump NA of 0.45. This LCF had pump absorption of 11dB/m at ~976nm. The LCF also had 2ρ=52.7μm and d/Λ=0.8. This gives a simulated effective area of 1548μm^2 at 1.05μm. The LCF has a rounded hexagonal shape and a flat-to-flat dimension of 254.2μm. The fiber was used to demonstrate amplification of 600ps pulses (with 600μJ pulse energy) to 1MW peak power.

Ytterbium double-clad all-glass LCFs with highly fluorine-doped silica as pump cladding were also demonstrated (see Figure 16) [37]. All-glass LCFs have no polymer in the pump path and have independent control of the fiber outer diameters and pump cladding dimensions. This, therefore, enables designs with smaller pump guides for higher pump absorption and, at the same time, with larger fiber diameters to minimize micro and macro bending effects, a much desired feature for large core fibers where intermodal coupling could be an issue due much increased mode density. Stress rods can also be added for PM LCFs (see Figure 16).

(a) (b)

Figure 16. An all glass active LCF (left) and an all glass PM active LCF with highly fluorine-doped pump cladding (right) [37].

The LCFs had a refractive index difference between the background and the low index feature of $\Delta n=1.2\times10^{-3}$. The non-PM LCF (see the left figure in Figure 16) had an inner layer $d/\Lambda=0.8$ and outer layer $d/\Lambda=0.7$. It had a 47μm core diameter, a rounded hexagon pump guide with a dimension of 238μm by 256μm, a pump NA of 0.28, pump absorption of ~12dB/m, and an outer diameter of ~538μm coated with standard high index coating, shown as the outermost layer in Figure 16(a). A 3.5m fiber coiled in 53cm diameter was used to demonstrate a slope efficiency of 75% in an amplifier [37]. A single stage gain of 33dB was demonstrated using this fiber. It was also used to directly amplify 15ps pulses to a peak power of ~1MW.

The PM LCF (see Figure 16(b)) had a core diameter of 80μm and had a fluorine-doped pump cladding, providing a pump NA of ~0.28. Low index features with an inner layer d/Λ of 0.8 and an outer layer d/Λ of

0.7 were used. This active PM LCF had a pump guide diameter of ~400μm (flat-to-flat), a fiber outer diameter of ~835μm, and was coated with standard acrylic coating. Pump absorption was estimated to be ~12dB/m. The mode field diameter was measured to be ~62μm. The fiber was used as an amplifier in a single coil 76cm in diameter with a length of straight section at each end, demonstrating a slope efficiency of ~74% and a maximum single-path gain in excess of 30dB [37]. It demonstrated direct amplification of 14.2ps pulses to 190kW peak power with pulse energy of 2.74μJ and negligible SPM spectral broadening. M^2 was measured to be below 1.35 for the entire output power range.

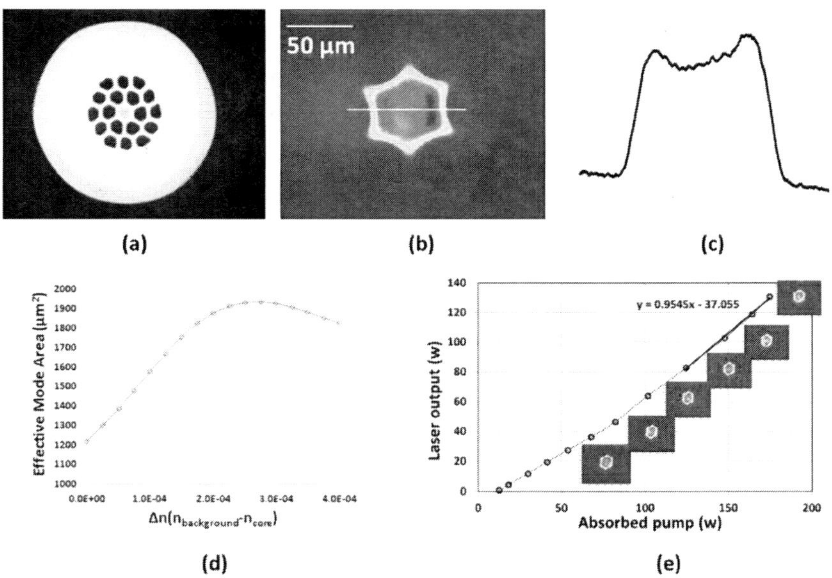

Figure 17. (a) The ytterbium-doped LCF with an index depression in the core center, (b) measured near field intensity of the guided mode, (c) measured mode intensity distribution, (d) simulated effective mode area versus the index depression, and (d) measured laser output and near field patters at various powers [40].

Recently, a flat-top mode has been demonstrated in an ytterbium-doped LCF with a ~50μm core by introducing an area ~30μm in diameter in the core center with a refractive index of $\sim 2\times10^{-4}$ lower than that of the background glass (Figure 17(a)) [40]. The flat-top mode (see Figure 17(b) and (c)) increased the effective mode area of the LCF from $\sim 1200\mu m^2$ to $\sim 1900\mu m^2$, a ~50% increase (see Figure 17(d)). The LCF also demonstrated near quantum-limited efficiency (see Figure 17(e)). Lasing wavelength was 1026nm and the pump was at 976nm.

4. HIGHER-ORDER-MODE FIBERS

The concept and demonstration of robust propagation of *higher-order-modes* (HOM) with large effective mode areas were first reported in 2006 [11]. It was argued that the stability of mode propagation in multimode fibers is critically dependent on the effective mode index difference between the propagating mode and its nearest neighbor anti-symmetric mode. This effective mode index difference is actually larger for the higher-order LP_{04}-LP_{07} modes than for the fundamental mode (see Figure 18). In addition, a *long period grating* (LPG), a fairly matured technology, can be used for broad band mode conversion to and from those higher-order LP0n modes.

Mode coupling first requires phase-matching. Perturbations on the fiber need to provide a vector which equals the difference in the propagation constants of the two modes involved. The larger the mode index difference between the two modes, the larger is the wave vector required for phase matching. A large wave vector implies high spatial frequency and small spatial period. For a mode index difference of 1.5×10^{-4}, the required spatial period of the perturbation at 1.55μm is ~10mm. Due to geometric constraints, the spatial frequency distribution of perturbations on fibers usually cuts off at a certain upper frequency limit. This would minimize mode coupling between modes with larger effective mode index differences. The second requirement for mode coupling is that the overlap integral among the two modes and the perturbation needs to be non-zero. This requires the perturbation to break up the mode orthogonality. Although it was not explicitly spelled out in [11], it was assumed that the perturbations on fibers are mostly anti-symmetric. In this case, mode coupling dominates between LP0n and its anti-symmetric counterpart LP1n modes in optical fibers.

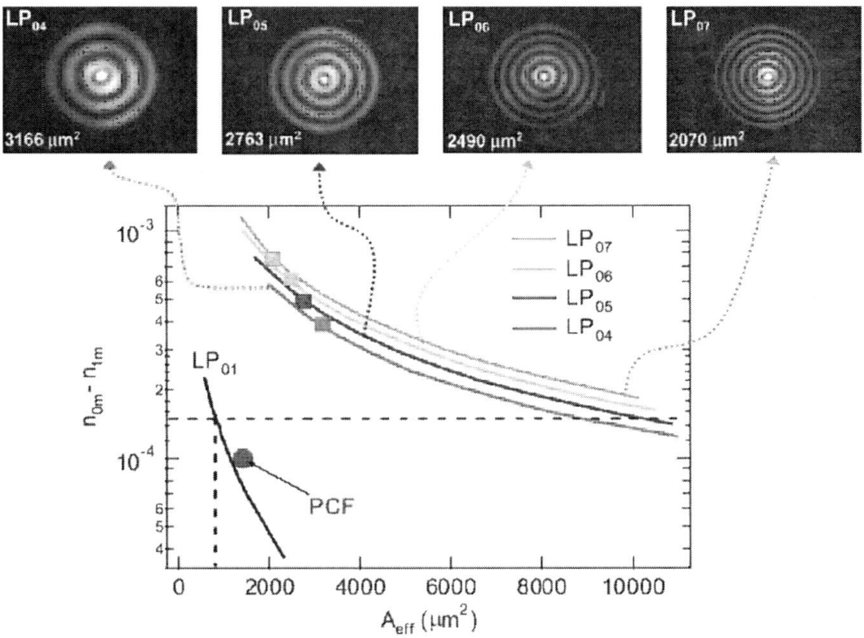

Figure 18. Effective index difference between nearest neighbor anti-symmetric mode versus A_{eff}. Top, near-field images of LPG-excited HOMs after >2m propagation with 7cm bend and with A_{eff} ranging from 2100 to 3200µm^2 [11].

The effective mode index difference for LP_{01}, LP_{04}, LP_{05}, LP_{06} and LP_{07} modes with their nearest neighbor anti-symmetric modes are shown versus effective mode area in Figure 18 [11]. The near-field images of the measured HOM modes are shown at the top of in Figure 18 and in Figure 19. It is clear that the effective mode index difference is significantly larger for the LP_{04}-LP_{07} modes than for the LP_{01} mode with the same effective mode area. The effective mode index difference also increases slightly for higher order LP_{0n} modes with the same effective mode area. For the same fiber, the effective mode area is actually larger for lower-mode-order LP_{0n} modes. The schematic of the proposed system is shown in Figure 20(a), where two identical LPGs are required for mode conversion from the LP_{01} mode to the LP_{07} mode and back to the LP_{01} mode. Robust propagation of the LP_{07} mode with an effective mode area of 2070μm^2 was demonstrated at 1600nm with the arrangement shown in Figure 19(c). The performance of the mode converter with high efficiency over broad bandwidth is shown in Figure 20(b). It was found that the LP_{07} mode suffered negligible bend loss at coil diameters down to 12cm. It was also found that modal stability increases with mode order. In a second paper [12], the effective mode areas of LP_{07} mode was simulated at various bend radii, showing stronger bend resistance than both LP_{03} and LP_{01} modes.

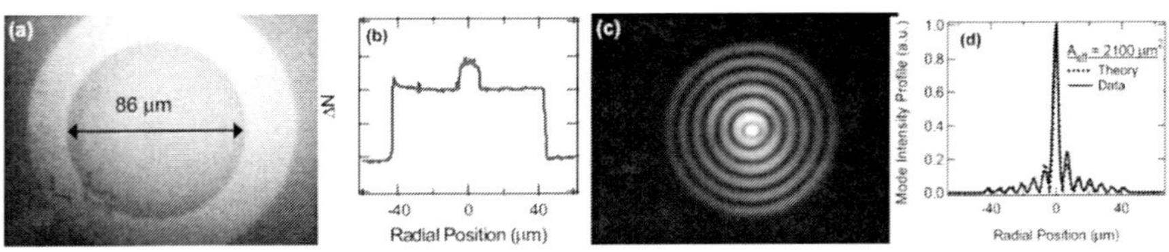

Figure 19. Characteristics of an HOM fiber. The horizontal and vertical scales of the images are identical. (a) Near-field image of a fiber facet, showing the 86μm inner cladding. (b) Refractive-index profile of the HOM fiber, with a core similar to an SMF with an 86μm inner cladding and a down-doped outer trench. (c) Near-field image of the LP_{07} mode at 1600 nm after 12m propagation with a 4.5cm radius bend. (d) Intensity line scan of (c) and theoretical profile: A_{eff}=~2100μm^2 [11].

Recently, an erbium-doped HOM amplifier was demonstrated [41]. A small inner core was designed to have a LP_{01} mode MFD of 9μm, allowing for effective excitation of the LP_{01} mode when the HOM fiber was spliced to a single-mode fiber. Higher-order modes expanded to occupy the outer core. The LP_{010} mode had an effective mode area of 2700μm². Both the inner and the outer core were doped with erbium with absorption of ~30dB/m at 1530nm. The amplifier was both seeded at 1564nm and pumped at 1480nm in the LP_{010} mode. The pump was a Raman fiber laser.

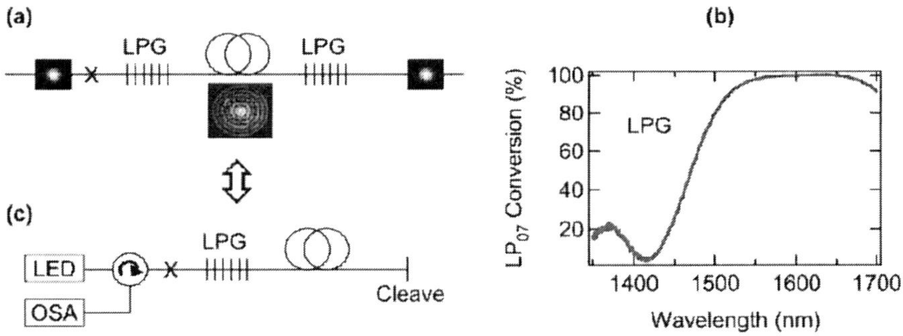

Figure 20. (a) Device schematic: light is coupled into and out of HOM with LPGs whose conversion efficiency being shown in (b). LPG with broadband coupling with efficiency >99% over 94 nm and with peak coupling efficiency as high as 99.93%. (c) Alternative schematic for characterizing HOM fibers: the cleave serves to fold the device propagation path so that the single LPG acts as both the input and the output LPG. X, splice; OSA, optical spectrum analyzer [11].

The difference in effective index between nearest neighbor anti-symmetric modes at λ=1564 nm as a function of their effective area is plotted as points in Figure 21(a) for the LP_{01} through LP_{010} modes in the fiber. The LP_{02} and LP_{03} modes have a large A_{eff}, but small mode spacing. As the mode order increases, mode spacing increases too, while A_{eff} decreases. Figure 21(b) shows the calculated intensity profiles at λ=1564nm for the LP_{01} and LP_{010} mode.

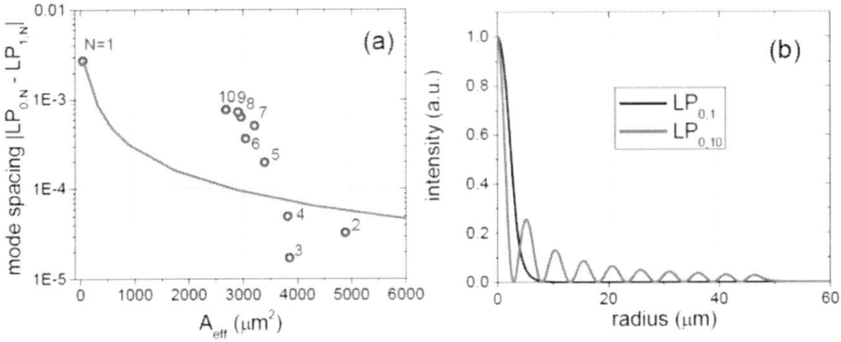

Figure 21. (a) Mode spacing as a function of effective area for the LP_{0n} modes in the HOM fiber (points) compared to a conventional LP_{01} step-index fiber with V=5 (solid curve). (b) Intensity profiles of the LP_{01} and LP_{010} modes. These calculations were done at a wavelength of 1564 nm.

A narrow line width, external cavity laser was amplified to 50mW and combined with the high-power, single mode Raman fiber laser at 1480 nm in a single-mode pump/signal combiner. The output of the pump/signal combiner was fusion-spliced to the HOM fiber. The length of the amplifier fiber after the LPG was 2.68 m. The measured slope efficiency at 1564nm was 43.2%. Over 20dB of gain was demonstrated by the amplifier.

5. CHRIALLY-COUPLED CORE FIBERS

A *Chirally-coupled-core* (CCC) fiber was first reported in 2007 [9]. The fiber had a large central core and a smaller side core wound around the central core in a helical fashion (Figure 22). The preform had two parallel cores and the fiber was spun during the draw to form the helical side core. In this first report, the central core had a diameter of 35μm and a NA of 0.07. The side core had a diameter of 12μm and NA of 0.09. Edge-to-edge core separation was 2μm. The helical pitch was 6.2mm. The fiber was measured to be single-mode at 1550nm over a short length of 25cm. It was multimode below 1500nm. The simulation predicted LP_{01} mode loss to be 0.3dB/m and all HOM loss to be >130dB/m for λ>1550nm. The fiber was also confirmed to be polarization-maintaining.

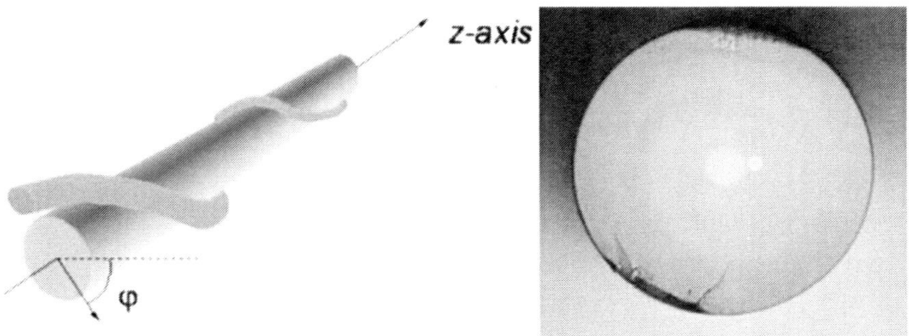

Figure 22. Structure of Chirally-coupled-core fiber [9].

The propagation of modes in the central core was affected by the coupling of modes between the central and side cores. The fiber was designed to operate where there was no fundamental mode coupling with modes in the side core. Higher-order modes in the central core were, however, coupled with the side core modes at the operating wavelength. The modes in the side core had high loss due to the tight bend from the helical arrangement. This led to high loss for the higher-order modes in the central core which were coupled to modes in the side core.

An ytterbium-doped CCC fiber was demonstrated in a subsequent paper [10]. The ytterbium-doped central core had a 33μm diameter and 0.06 NA. The undoped side core had a 16μm diameter and 0.1 NA. The helical pitch was 7.4mm and the edge-to-edge core separation was 4μm. The low index coating provided a pump NA of 0.47. The pump guide had a 250μm diameter. The measured pump absorption was 2dB/m at

915nm. The fiber demonstrated 75% slope efficiency at 1066nm in a laser configuration.

In a more recent paper [42], a more detailed theoretical analysis of quasi-phase-matching (QPM) assisted by spin and orbital angular momentum was given. For two LP modes LP_{l1m1} and LP_{l2m2}, QPM is achieved when

$$\beta_{l1m1} - \beta_{l2m2}\sqrt{1 + K^2R^2} - \Delta mK = 0 \quad (2)$$

where β_{l1m1} and β_{l1m2} are the respective propagation constants of the two modes; $K=2\pi/\Lambda$; and Λ is helical pitch. $\Delta m=\Delta l+\Delta s$; $\Delta l=\pm l1\pm l2$; $\Delta s=-2,-1, 0, 1, 2$. The loss versus wavelength for the LP_{11} and LP_{21} modes in the side core is plotted in Figure 23(a) showing high loss for these modes in certain wavelength regimes. The mode indices of the LP_{11} and LP_{21} modes of the side core and LP_{01} mode of the central core are plotted in Figure 23(b). It can be seen clearly that the LP_{11} mode and LP_{21} modes of the side core naturally phase-match to the LP_{01} mode of the central core at ~1.22μm and ~0.81μm respectively. The QPM by angular momentum from the helical side core extends the phase matching to a number of other wavelengths determined by equation 2. These phase matching wavelengths are plotted in Figure 23(b) as red dotted vertical lines for the LP_{11} mode of the side core and the LP_{01} mode of central core coupling and blue dotted vertical lines for the LP_{21} mode of side core and LP_{01} mode of central core coupling. The measured transmission of the central core is plotted in Figure 23(c). The analysis of the CCC fiber can be simplified significantly in a helical coordination. The Maxwell equations keep the same form in the new coordination system, but the tenor form of permittivity and permeability needs to be transformed [42]. The resulting tenor does not have any z-dependence, which significantly simplifies the analysis. The simulated loss of various modes is shown in Figure 23(d), demonstrating that the narrow peaks in the transmission arise from coupling between the LP_{01} mode in the central core and the LP_{11} and LP_{21} modes in the side core.

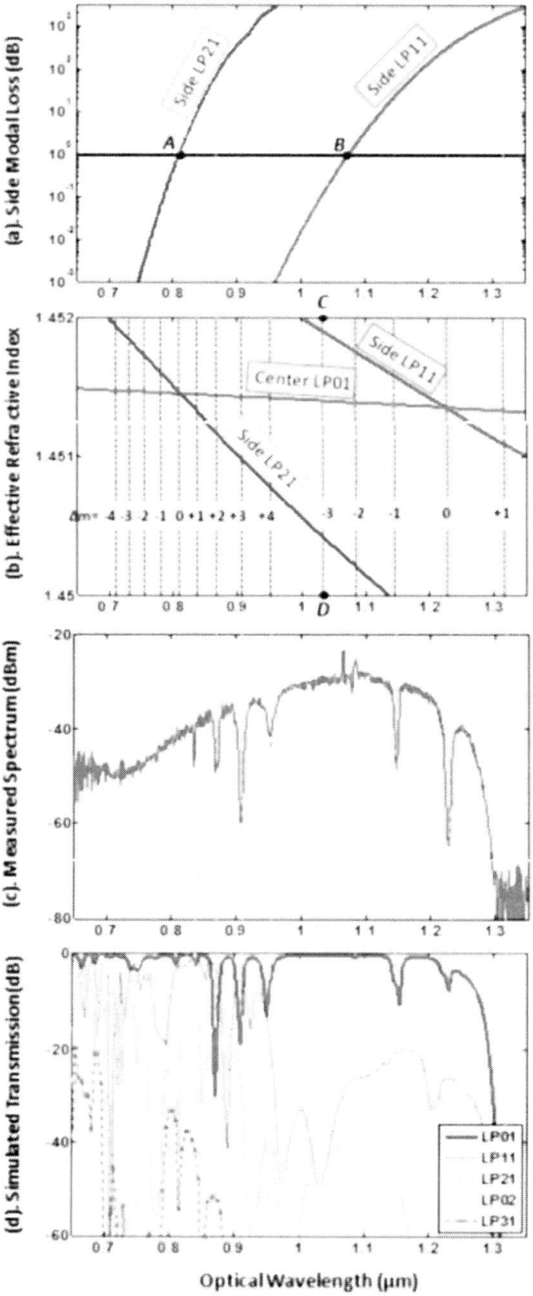
(a). Side Modal Loss (dB)
Side LP21
Side LP11
A
B
C
D
(b). Effective Refractive Index
Center LP01
Side LP11
Side LP21
Δm= -4 -3 -2 -1 0 +1 +2 +3 +4
-3 -2 -1 0 +1
(c). Measured Spectrum (dBm)
(d). Simulated Transmission (dB)
LP01
LP11
LP21
LP02
LP31
Optical Wavelength (μm)

Figure 23. Calculation and measurement of quasi-phase-matching (QPM) for 1.5m-long CCC fiber sample. (a) Calculated loss for side-core LP_{11} and LP_{21} modes as a function of wavelength (top). (b) Calculated refractive indices of interacting modes and calculated QPM resonance positions. (c) Transmission of central core. (d) Simulated transmission of central core.

The high loss of the LP_{01} mode in the central core at $\lambda>1.3\mu m$ was not explained in the paper. It may be due to the angular momentum assisted coupling between LP_{01} mode in the central core and LP_{01} mode in the side core. This long wavelength cut-off has been used for the suppression of stimulated Raman scattering in fibers [43].

The higher loss for the higher-order-modes in the central core between 1-1.1μm in Figure 23(d) was not clearly explained either. For the LP_{11} mode in the central core, one possible reason for its high loss is coupling to LP_{21} mode of the side core through angular momentum-assisted QPM. Recently, a 60μm core CCC fiber was also demonstrated [44].

6. PHOTONIC BANDGAP FIBERS

The guidance of light in a defect in a photonic bandgap fiber is based on fundamentally different principles compared to conventional optical fibers. In a conventional optical fiber, the core has a higher refractive index than that of the cladding. Light is guided through total internal reflection at the core-and-cladding boundary. In a photonic bandgap fiber, the photonic crystal cladding is designed to have photonic bandgaps, where there is an absence of modes propagating in the direction parallel to the fiber axis. Once a defect core is created within such a photonic crystal cladding, light is trapped to propagate only in modes of the defect core as it is forbidden to propagate in the cladding. This leads to significantly lower waveguide loss for the modes in the defect core within the cladding photonic bandgaps. This waveguide loss becomes zero when the photonic crystal cladding is infinite, but there is always a finite loss for practical fibers with finite cladding. This waveguide loss can, however, be significantly lowered by increasing the number of cladding layers.

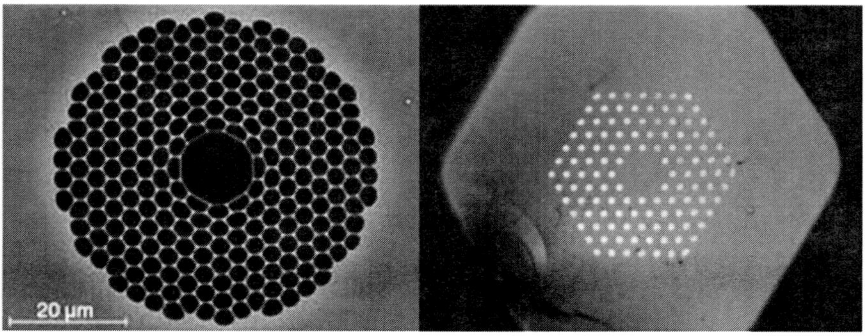

Figure 24. Images of a hollow-core photonic bandgap fiber (left) and an all-solid photonic bandgap fiber (right).

One unique feature of photonic bandgap fibers is that the defect core always has a lower refractive index than that of the higher refractive index material in the cladding. This leads to the possibility of hollow-core photonic bandgap fibers (see Figure 24, left), where light is guided mostly in the air in the hollow-core of the fiber. These fibers have extremely low nonlinearities and are well suited for high power laser delivery. The second type of photonic bandgap fibers are made entirely of glass (see Figure 24, right). These all-solid photonic bandgap fibers have a cladding which consists of a background glass and nodes of slightly higher refractive index. The much lower refractive index contrast (typically just a few percent) in the photonic crystal cladding of the all-solid photonic bandgap fibers still allows photonic bandgaps for paraxial propagation. Another important feature of photonic bandgap fibers is that their transmission is highly wavelength-dependent, i.e. low loss is possible only within the photonic bandgaps of the photonic crystal cladding. This distributive spectral filtering along a fiber can be very useful for range of potential applications including use in fiber lasers at low gain regimes to minimize gain competition, the suppression of stimulated Raman scattering, etc.

Birks et al. clearly explained the origin of the modes in the photonic crystal cladding of an all-solid photonic bandgap fiber in [45]. The density of states (DOS) of modes in the photonic crystal cladding is represented by the shades of grey in Figure 25 [45]. The refractive index of the background glass is represented by the black horizontal line. It is the cut-off line for the modes guided in the cladding nodes. The bandgap regime is represented by the red area. The fundamental mode of the defect core is illustrated by the yellow lines. The modes in the photonic crystal cladding clearly originate from the guided modes of the nodes, which are also labeled at the top of the figure. Above the background index, the modes form relatively narrow bands. Below the background index, the bands of modes significantly broaden, due to strong coupling among nodes as a result of light becoming more spread into the background glass below cut-off. The guidance property can be easily understood once it is recognized that there is a simple analogue to conventional optical fibers [46]. The core index is simply the index of the background glass and the equivalent cladding index is the upper boundary of the photonic bandgap. The effective index of the modes in the defect core (see yellow lines in Figure 25) falls between the core and cladding indices as in conventional optical fibers.

One interesting application of ytterbium-doped all-solid photonic bandgap fibers is for the lasing of ytterbium at the long wavelength regime of 1150-1200nm [47]. It is normally difficult to lase in ytterbium-doped fibers at these extremely long wavelengths due to strong gain competition from the short wavelength regime of 1030-1070nm. The distributive spectral filtering in a photonic bandgap fiber can be used to minimize gain at short wavelengths, leading to efficient high power lasers in the extremely long wavelength regime.

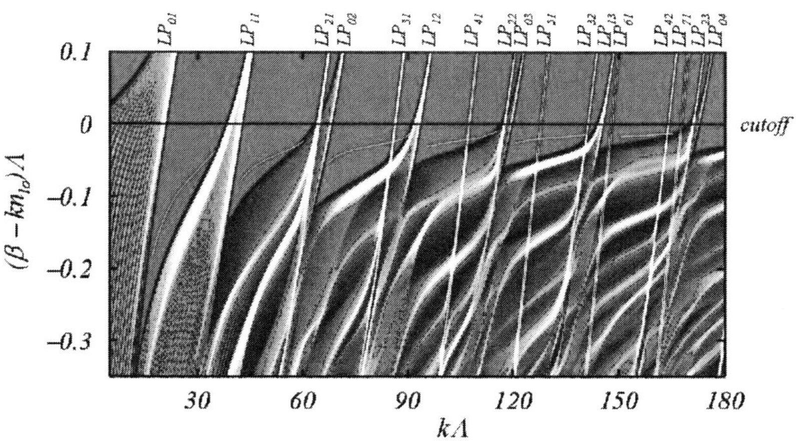

Figure 25. Plots of band structure for an example bandgap fiber. The bandgaps are shown in red. The node modes from which the bands arise are labeled along the top. Density of states (DOS) calculated using the plane-wave method, with light grey corresponding to high DOS. The yellow curve is the "fundamental" core-guided mode [45].

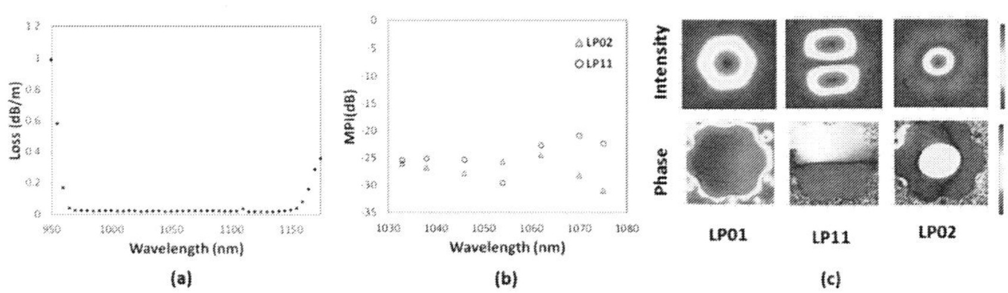

Figure 26. (a) Transmission of the 55μm core all-solid photonic bandgap fiber, (b) measured HOM contents with a S^2 method in a 5m fiber coiled at 70cm diameter, and (c) measured mode images [51].

In all-solid photonic bandgap fibers, a mode is guided only when it falls within the photonic bandgap of the cladding lattice. Guidance can therefore be highly mode-dependent. This provides great potential for creating designs that support only the fundamental mode, i.e. selective mode guidance. Mode area scaling to 20µm mode field diameter using all-solid photonic bandgap fibers was reported in [48]. Recently, all-solid photonic bandgap fibers with up to ~700µm² effective mode area have been demonstrated operating in the first bandgap [49, 50]. More recently, a fiber with a core diameter of ~55µm and an effective mode area of 920µm² at coil diameter of 50cm was demonstrated (see right figure in Figure 24). The fiber operates in the third bandgap with transmission shown in Figure 24(6). At 70cm coil diameter, a 5m fiber showed HOM contents below 25dB (see Figure 26(b) and (c)). At the design coil diameter of 50cm, HOM content was below-30dB [51].

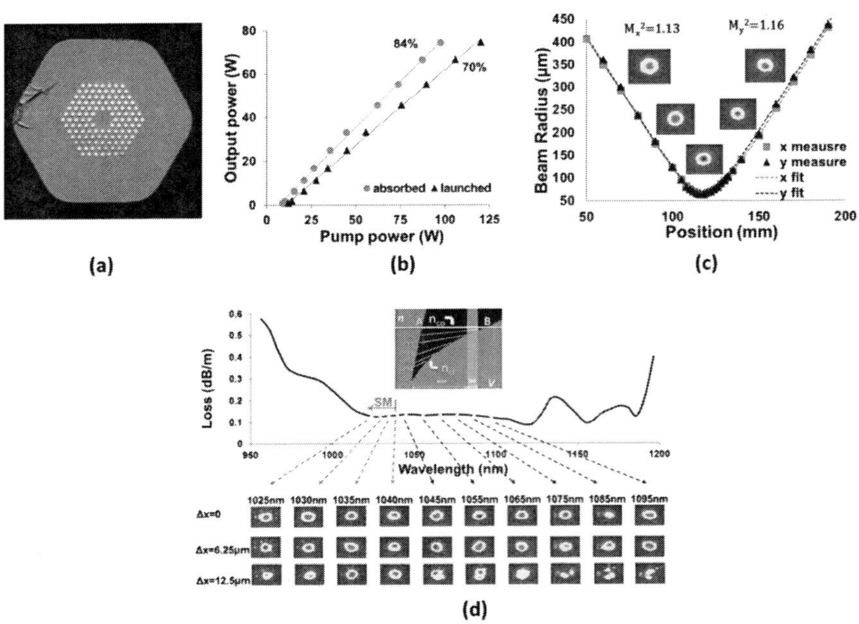

Figure 27. (a) Ytterbium-doped ~50µm core all-solid photonic bandgap fiber, (b) Laser efficiency, (c) M² measurement and (d) mode at various wavelengths across the bandgap [52].

Recently, an ytterbium-doped all-solid photonic bandgap fiber with ~50μm core diameter has also been demonstrated (Figure 27(a)) [52]. The fiber demonstrated high efficiency and excellent mode quality (see Figure 27(b) and (c)). The fiber also demonstrated robust single-mode behavior near the short wavelength edge of the bandgap by monitoring the output when moving away from the optimum launch condition (see Figure 27(d)). This is exactly expected from the dispersive nature of the photonic bandgap cladding.

7. CONCLUSIONS

In this chapter, we have briefly introduced a number of emerging fiber technologies for mode area scaling for fiber lasers. These new technologies are critical for the further power scaling of fiber lasers. The basic principles of these technologies were introduced and the latest developments were discussed. This is still a very active area of research. With further development, there is great potential to significantly improve throughput and expand the capabilities of fiber lasers for use in manufacturing, as well as to meet future defense needs.

REFERENCES

1. D. Taverner, D.J. Richardson, L. Dong, J.E. Caplen, K. Williams, R.V. Penty, "158μJ pulses from a single transverse mode, large mode-area EDFA", Optics Letters, 22, 378-380 (1997).

2. H.L. Offerhaus, N.G. Broderick, D.J. Richardson, R. Sammut, J. Caplen and L. Dong, "High-energy single-transverse-mode Q-switched fiber laser based on a multimode large-mode-area erbium-doped fiber", Optics Letters, 23, 1683-1685 (1998).

3. C. Knight, T. A. Birks, P. St. J. Russell, and D. M. Atkin, "All-silica single-mode optical fiber with photonic crystal cladding," Opt. Lett. 21, 1547-1549 (1996).

4. T.A. Birks, J.C. Knight and P.St.J. Russell, "Endlessly single-mode photonic crystal fiber," Opt Lett. 22 961-963 (1997).

5. J.C. Knight, T.A. Birks, R.F. Cregan, P.St.J. Russell and J.P. de Sandro, "Large mode area photonic crystal fibre," Elect. Lett. 34, 1347-1348 (1998).

6. C.D. Brooks and F. Teodoro, "Multi-MW peak-power, single-transverse-mode operation of a 100μm core diameter, Yb-doped rodlike photonic crystal fiber amplifier," Appl. Phys. 89, 111119 (2006).

7. W.S. Wong, X. Peng, J.M. McLaughlin and L. Dong, "Breaking the limit of maximum effective area for robust single-mode propagation in optical fibers", Optics Letters, 30, 2855-2857 (2005).

8. L. Dong, X. Peng and J. Li, "Leakage channel optical fibers with large effective area", Journal of Optical Society of America B, 24, 1689-1697 (2007).

9. C.H. Liu, G. Chang, N. Litchinitser, A. Galvanauskas, D. Guertin, N. Jacobson and K. Tankala, "Effective single-mode chirally-coupled core fiber," Advanced Solid State Photonics, paper ME2, 2007.

10. A. Galvanauskas, M.C. Swan, and C.H. Liu, "Effectively-Single-Mode Large Core Passive and Active Fibers with Chirally-Coupled-Core structures," Conference on Lasers and Electro-Optics, paper CMB1, 2008.

11. S. Ramachandran, J. W. Nicholson, S. Ghalmi, M. F. Yan, P. Wisk, E. Monberg, and F. V. Dimarcello, "Light propagation with ultralarge modal areas in optical fibers," Opt. Lett. 31, 1797-1799 (2006).

12. J.M. Fini and S. Ramachandran, "Natural bend-distortion immunity of higher-order-mode large-mode-area fibers" Opt. Lett. 32, 748-750 (2007).

13. M. Koshiba and K. Saitoh, "Applicability of classical optical fiber theories to holey fibers," Opt. Lett. 29, 1739-1740 (2004).

14. K. Saitoh, Y. Tsuchida, and M. Koshiba, "Endlessly single-mode holey fibers: the influence of core design," Opt. Express 13, 10833-10839 (2005).

15. B. T. Kuhlmey, R. C. Mcphedran, and C. M. de sterke, "Modal cutoff in microstructured optical fiber," Opt. Lett. 27, 1684-1686 (2002).

16. J. Zhou, K. Tajima, K. Nakajima, K. Kurokawa, C. Fukai, T. Matsui, and I. Sankawa, "Progress on low loss photonic crystal fibers," Optical Fiber Technology 11 101–110 (2005).

17. T. P. White, B. T. Kulmey, R. C. McPhedran, D. Maystre, G. Renversez, C. Martijn de Sterke, and L. C. Botten, "Multipole method for microstructured optical fibers. I. Formulation," J. Opt. Soc. Am. B 19, 2322-2330 (2002).

18. B.T. Kuhlmey, T.P. White, G. Renversez, D. Maystre, L.C. Botten, C. M. de Sterke, and R. C. McPhedran, "Multipole method for microstructured optical fibers. II. Implementation and results," J. Opt. Soc. Am. B 19, 2331-2340 (2002).

19. W.J. Wadsworth, J.C. Knight and P.St.J. Russell, "Large mode area photonic crystal fiber laser," Conference on Lasers and Electro-Optics, CWC1, 2001.

20. W.J. Wadsworth, R.M. Percival, G. Bouwmans, J.C. Knight and P.St.J. Russell, "High power air-clad area photonic crystal fiber laser," Opt. Express 11, 48-53 (2003).

21. D.J. DiGiovanni and R.S. Windeler: "Article comprising an airclad optical fiber," US patent 5907652 (1999).

22. V. A. Kozlov, J. Hernández-Cordero, R. L. Shubochkin, A. L. G. Carter, and T. F. Morse, "Silica–Air Double-Clad Optical Fiber," IEEE Photonics Technology Letters 12, 1007-1009 (2000).

23. J.K. Sahu, C.C. Renaud, K. Furusawa, R. Selvas, J.A. Alvarez-Chavez, D.J. Richardson and J. Nilsson, "Jacketed air-clad cladding pumped ytterbium-doped fibre laser with wide turning range," Elect. Lett. 37, 1116-1117 (2001).

24. K. Furusawa, A. Malinowski, J. H. V. Price, T. M. Monro, J. K. Sahu, J. Nilsson and D. J. Richardson, "Cladding pumped Ytterbium-doped fiber laser with holey inner and outer cladding," Opt. Express 9, 714-720 (2001).

25. W.J. Wadsworth, R.M. Percival, G. Bouwmans, J.C. Knight, T.A. Birks, T.D. Hedley, and P.St.J. Russell, "Very high numerical aperture fibers," IEEE Photonic Technology Letters 16, 843-845 (2004).

26. J. Limpert, T. Schreiber, S. Nolte, H. Zellmer, A. Tünnermann, R. Iliew, F. Lederer, J. Broeng, G. Vienne, A. Petersson, and C. Jakobsen, "High-power air-clad large-mode-area photonic crystal fiber laser," Opt. Express 11, 818-823 (2003).

27. J. Limpert, T. Schreiber, A. Liem, S. Nolte, H. Zellmer, T. Peschel, V. Guyenot, and A. Tünnermann, "Thermo-optical properties of air-clad photonic crystal fiber lasers in high power operation," Opt. Express 11, 2982-2990 (2003).

28. J. Limpert, A. Liem, M. Reich, T. Schreiber, S. Nolte, H. Zellmer, A. Tünnermann, J. Broeng, A. Petersson, and C. Jakobsen, "Low-nonlinearity single-transverse-mode ytterbium-doped photonic crystal fiber amplifier," Opt. Express 12, 1313-1319 (2004).

29. J. Limpert, N. Deguil-Robin, I. Manek-Hönninger, F. Salin, F. Röser, A. Liem, T. Schreiber, S. Nolte, H. Zellmer, A. Tünnermann, J. Broeng, A. Petersson, and C. Jakobsen, "High-power rod-type photonic crystal fiber laser," Opt. Express 13, 1055-1058 (2005).

30. J. Limpert, O. Schmidt, J. Rothhardt, F. Röser, T. Schreiber, and A. Tünnermann, S. Ermeneux, P. Yvernault, and F. Salin, "Extended single-mode photonic crystal fiber lasers," Opt. Express 14, 2715-2720 (2006).

31. T. Schreiber, F. Röser, O. Schmidt, J. Limpert, R. Iliew, F. Lederer, A. Petersson, C. Jacobsen, K. P. Hansen, J. Broeng, and A. Tünnermann, "Stress-induced single-polarization single-transverse mode photonic crystal fiber with low nonlinearity," Opt. Express 13, 7622-7630 (2005).

32. O. Schmidt, J. Rothhardt, T. Eidam, F. Röser, J. Limpert, A. Tünnermann, K.P. Hansen, C. Jakobsen, and J. Broeng, "Single-polarization ultra-large-mode-area Yb-doped photonic crystal fiber," Opt. Express 16, 3918-3923 (2008).

33. L. Dong, J. Li and X. Peng, "Bend resistant fundamental mode operation in ytterbium-doped leakage channel fibers with effective areas up to 3160μm2", Optics Express, 14, 11512-11519 (2006).

34. X. Peng and L. Dong, "Fundamental mode operation in polarization-maintaining ytterbium-doped fiber with an effective area of 1400μm2", Optics Letters, 32, 358-360 (2006).

35. L. Dong, T.W. Wu, H.A. McKay, L. Fu, J. Li and H.G. Winful, " All-glass large core leakage channel fibers, " IEEE Journal of Selected Topics in Quantum Electronics, 15, invited paper, 47-53(2009).

36. L. Dong, H.A. Mckay, A. Marcinkevicius, L. Fu, J. Li, B.K. Thomas, and M.E. Fermann, " Extending Effective Area of Fundamental Mode in Optical Fibers, " IEEE Journal of Lightwave Technology, 27, invited paper, 1565-1570(2009).

37. L. Dong, H.A. Mckay, L. Fu, M. Ohta, A. Marcinkevicius, S. Suzuki, and M.E. Fermann, "Ytterbium-doped all glass leakage channel fibers with highly fluorine-doped silica pump cladding, " Optical Express, 17, 8962-8969(2009).

38. G. Gu, F. Kong, T.W. Hawkins, P. Foy, K. Wei, B. Samson, and L. Dong, "Impact of fiber outer boundaries on leaky mode losses in leakage channel fibers," Optics Express 21, 24039-24048 (2013).

39. J. W. Nicholson, A. D. Yablon, S. Ramachandran, and S. Ghalmi, "Spatially and spectrally resolved imaging of modal contents in large-mode-area fibers," Opt. Express 16, 7233-7243(2008).

40. F. Kong, G. Gu, T.W. Hawkins, J. Parsons, M. Jones, C. Dunn, M.T. Kalichevsky-Dong, K. Wei, B. Samson, and L. Dong, "Flat-top mode from a 50 μm-core Yb-doped leakage channel fiber," Optics Express 21, 32371-32376 (2013).

41. J.W. Nicholson, J.M. Fini, A.M. DeSantolo, E. Monberg, F. DiMarcello, J. Fleming, C. Headley, D.J. DiGiovanni, S. Ghalmi, and S. Ramachandran, "A higher-order-mode Erbium-doped-fiber amplifier," Opt. Express 18, 17651-17657 (2010).

42. X. Ma, C.H. Liu, G. Chang, and A. Galvanauskas, "Angular-momentum coupled optical waves in chirally-coupled-core fibers," Opt. Express 19, 26515-26528 (2011).

43. X. Ma, I,N Hu, and A. Galvanauskas, "Propagation-length independent SRS threshold in chirally-coupled-core fibers," Opt. Express 19, 22575-22581 (2011).

44. X. Ma, A. Kaplan, and A. Galvanauskas, "Experimental characterization of robust single-mode operation of 50μm and 60μm core chirally coupled core optical fibers," PhotonicsWest, paper 8237-59, 2012.

45. T. A. Birks, G. J. Pearce, D. M. Bird, "Approximate band structure calculation for photonic bandgap fibers," Opt. Express 14, 9483-9490 (2006).

46. M.J.F. Digonnet, H.K. Kim, G.S. Kino, and S. Fan, "Understanding Air-Core Photonic-Bandgap Fibers: Analogy to Conventional Fibers," Journal of Lightwave Technology 23, 4146-4177 (2005).

47. A. Shirakawa, H. Maruyama, K. Ueda, C. B. Olausson, J. K. Lyngsø, and J. Broeng, "High-power Yb-doped photonic bandgap fiber amplifier at 1150-1200 nm," Opt. Express 17(2), 447–454 (2009).

48. O. N. Egorova, S. L. Semjonov, A. F. Kosolapov, A. N. Denisov, A. D. Pryamikov, D. A. Gaponov, A. S. Biriukov, E. M. Dianov, M. Y. Salganskii, V. F. Khopin, M. V. Yashkov, A. N. Gurianov, and D. V. Kuksenkov, "Single-mode all-silica photonic bandgap fiber with 20-micron mode-field diameter," Opt. Express 16, 11735–11740 (2008).

49. M. Kashiwagi, K. Saitoh, K. Takenaga, S. Tanigawa, S. Matsuo, and M. Fujimaki, "Low bending loss and effectively single-mode all-solid photonic bandgap fiber with an effective area of 650μm2.," Opt. Lett. 37, 1292–1294 (2012).

50. M. Kashiwagi, K. Saitoh, K. Takenaga, S. Tanigawa, S. Matsuo, and M. Fujimaki, "Effectively single-mode allsolid photonic bandgap fiber with large effective area and low bending loss for compact high-power all-fiber lasers," Opt. Express 20, 15061–15070 (2012).

51. F. Kong, K. Saitoh, D. Mcclane,T. Hawkins, P. Foy, G.C. Gu, and L. Dong, "Mode Area Scaling with All-solid Photonic Bandgap Fibers,", Optics Express 20, 26363-26372 (2012).

52. G. Gu, F. Kong, T.W. Hawkins, J. Parsons, M. Jones, C. Dunn, M.T. Kalichevsky-Dong, K. Saitoh, and L. Dong, "Ytterbium-doped large-mode-area all-solid photonic bandgap fiber lasers," Optics Express 22, 13962-13968(2014).

CHAPTER 2

Dispersion Compensating Fibres for Fibre Optic Telecommunication Systems

Michal Lucki[1] and Tomas Zeman[1]

[1] Department of Telecommunication Engineering, Faculty of Electrical Engineering, Czech Technical University in Prague, Prague, Czech Republic

1. INTRODUCTION

Chromatic dispersion (CD) of optical pulses in an optical fibre influences their width. It refers to changes in propagation of particular frequency components contained in optical pulses causing extension of optical pulses. The fibre reach is then significantly limited, unless signal regenerators are used. Because the pulse spread degrades optical systems, it is necessary to prevent its origination or to eliminate its results. CD ought to be suppressed or one shall prevent its origination by specialty optical fibres, including both conventional and microstructured optical fibres (MOF). To keep the pulse length nearly constant, it is possible to use Dispersion Compensating Fibre (DCF) that shall be installed at the regenerators that provide amplification, renewal of timing and pulse duration. DCFs have negative dispersion parameter; CD accumulating between regenerators is then suppressed. Consequently, bit error rate could be improved and the fibre reach could be extended [1].

Photonic Crystal Fibres (PCF) could be a suitable solution for the problem of CD in high-speed transmission systems, especially those using Wavelength Division Multiplexing (WDM) to increase the symbol rate [2].

We systematize design approaches for specialty dispersion-tailored optical fibres in order to offer a guideline to flexibly design optical fibres used in telecommunications whose optimized CD is a key property. Last but not least, we present selected fibre designs prepared in last few years, including CD plots and information about structural parameters.

The knowledge about how to design fibres could be useful to design optical fibres, such as for example a submicron flat CD compensating fibre. In addition, fluoride compounds, as compared with silica glass, exhibit higher effective refractive index, the wider spectrum of working wavelengths (λ) or lower insertion losses. Additives using these materials are promising for extending the application towards infrared region, where fluoride glasses are usually transparent.

2. FIBRES WITH OPTIMIZED DISPERSION [3]

DCFs could regenerate signals that are spread as a consequence of CD. This practically means that the bit error rate at the receiver's side could be improved or the spaces between adjacent symbols could be reduced. As a result, potential bit rate could be increased. One of the approaches how to deal with CD is to use zero CD fibres, offering near zero CD at the operating λs. (It shouldn't be exactly zero because of Four Wave Mixing problem occurring when propagation is with zero dispersion and the phases of all the frequency components are matched). Another approach is to use already mentioned DCFs at the signal regenerators. CD tailoring fibres could work at some λ, e.g. at 1.55 μm, where they could have negative value of CD being few hundred picoseconds (ps/nm/km), which expresses the delay between the slowest and fastest frequency component measured at the distance of 1 km, assuming the source of radiation emitting the spectrum of 1 nm. Sample compensating fibre was published in Refs. [4] and [5]. Very low CD parameter is observed whose properties could benefit in compensation of CD accumulating in some optical network. An unsolved problem seems to be relatively narrow window of λs. Such a fibre could then compensate CD in just one λ channel of a WDM system [3].

Resultant CD is a balance between waveguide and material component. In order to obtain flat CD, the evolution of waveguide dispersion has to be exactly opposite to the one of material dispersion, which means that both ought to be optimized at each channel step by step. The situation is simple when material dispersion is linear upon λ. Then, one could employ an algorithmic approach with iterations to calculate and precisely adjust the CD at each λ. A very important assumption is that the balance between both dispersion components could be a very useful tool in designing fibres for CD tailoring by optimizing material dispersion (through the material properties) and waveguide dispersion (as a result of optimization of the shape of the waveguide) [3] [6].

2.1. Dispersion compensating fibres [2]

When a compensating fibre works at some λ, optical symbols ought to be transmitted at small spectrum of light waves, without multiplying the bit rate in many channels. Such narrowband fibres can have very low value of CD that is possible without using additives (e.g. germanium dioxide). In Ref. [7], it could be found that CD of -18 ns/nm/km is possible. Other works concentrate rather on the optimization of the structural parameters of MOFs to obtain flat CD diagram. Microstructured fibres offer much greater flexibility because geometry could be optimized not only through the core and cladding diameter, but also by changing the air filling fraction, and the lattice pitch including its arrangements. In Ref. [8], flat CD over all the telecom bands is shown [2].

Index guiding MOFs (the one without the central inclusion, having solid core) offer high flexibility in CD compensating fibres design. For example microstructured based DCFs could have dual cores [9] or they could be doped [10]. Negative value of CD parameter in DCFs with dual core structure could also be combined with the idea of making some rings with smaller holes [7][10]. Such a DCF with low CD parameter in broad telecom band is shown in Ref. [11], in addition, a square lattice (with the defining angle of 90°) was proposed. The performance of DCFs with the concentric cores could be increased by using germanium dioxide in the core [12]. The relation between the amount of additives and resultant CD is shown in Ref. [13]. Some other substrates could be used to lower the effective index (n_{eff}), such as for example fluorides [14]. On the contrary, germanium dioxide raises n_{eff}. The key feature is not to raise or lower a certain index, but to create large index contrast between the fibre's core index and its n_{eff} of the cladding [2].

Fibres with flat CD, i.e. those having low value of CD in many λ channels concurrently are often considered as wideband fibres, but they can't be used to compensate CD in all λ channels at once, because accumulated CD in each WDM channel is different. They are rather suitable to compensate CD at one channel, which can be selected from the wide range of λs that are compatible with the used DCF.Real wideband fibres should compensate CD at every λ channel at the same time. They must have CD exactly opposite to dispersion in each channel. In Ref. [15], it is shown that larger lattice pitch in the 1st ring of holes is responsible for the CD slope that must be optimized for this purpose. In addition, the hole radius in the 1st ring should be larger to enhance dispersion. CD slope property is studied in Ref. [16] in the context of wavelength division multiplexing. Another slope compensating DCF is shown in Refs. [2] and [17].

2.2. Zero dispersion wavelengths in DCFs

Because a great part of MOFs have parabolic wavelength evolution of CD and the minimum CD at low values, those fibres can have two zero-CD wavelengths (ZDW). The 1st ZDW is the one utilized at visible wavelengths, whereas the longer ZDW is used at telecom wavelengths [18]. Such Highly Nonlinear PCF (HNPCF) could exhibit positive slope at 1st ZDW and negative slope at 2nd ZDW, as shown in Ref. [19]; the 1st ZDW was at 900 nm, the 2nd one at the λ of 1.6 μm.

The high index difference between the air-filled microstructure and pure or doped silica core enables tight mode confinement resulting in a low effective area and, thereby, a non-conventional behaviour of CD. Modifying the periodic cladding (i.e., hole-sizes, lattice pitch), the waveguide dispersion and the origination of ZDWs are influenced, keeping the fibre still in single-mode operation regime [20]. Then, the strong wavelength dependence in the characteristics of the fibre will be used to determine either huge CD with large slope or nearly-zero flat CD.

The bandwidth, at which MOFs are designed to have zero CD, could be divided into three categories: working in the region between 0.55 μm and 1 μm, where 1st zero-CD point could be determined; another region between 1 μm and 1.2 μm, where 1st or 2nd ZDW could be found. The 3rd region, starting at 1.2 μm and going up to longer λs, is used to create the 2nd zero CD and is not available for the 1st zero CD. An overview of the obtainable ZDWs as a function of MOF's structure was shown in Ref. [21].

Both core size and the radius of holes exhibit significant influence on the location of both ZDWs. Increasing core size tunes the ZDW to longer λ. The increase in air percentage in the cladding could extend the origination of the 1st ZDW at longer wavelengths and the 2nd ZDW at shorter λs (i.e. less negative CD is obtained for larger air-filling fraction).

2.3. Dispersion flattened fibres and wideband fibres [2][3]

Some compensating fibres could work at short spectrum of λs. When a compensating fibre works at a certain λ, optical pulses can be transmitted using one or – in general – low amount of WDM channels. As it has already been mentioned above, MOFs are suitable for designing the compensating fibres because they allow huge index contrast and offer many parameters to be optimized (core size, hole radius, lattice pitch, amount of rings of holes) in order to optimize waveguide dispersion [2].

Currently, flat CD fibres could for example be referred to as near zero CD transmission medium. In Ref. [22], a DCF made of pure (undoped) material is shown. Advanced DCFs with flat CD property over broad spectrum of working λs could be found in Ref. [23]. An interesting fibre with CD parameter close to zero with CD fluctuations less than 0.5 ps in S to L bands is shown in Ref. [24]. Results were obtained by careful optimization of holes in particular rings. To reduce losses, it is often required avoiding doping the MOF's core. Instead, one could consider using octagonal [25] and decagonal [26] lattice, or more sophisticated lattice arrangements [2].

Recent analyses concentrate on the low, flat CD over telecommunication λs for any telecom wavelength compensation. The value of CD and the width of the working range is a compromise. A DSF could have large negative CD value, but for medium-wide wavelength range, or, acceptable CD parameter, but designed for very broad spectrum of λ. Currently, the strongest demand is to design DCFs mainly for C and L bands, where modern WDM systems can work.

The idea how to obtain flat CD properties is to tune the 1st ZDW to shorter λs and the 2nd ZDW to longer λs, having little negative CD over the whole bandwidth at the same time. Practically, predicted CD diagram is a wide parabola. Considered properties are related to the diameter of the core and the radius of the holes in particular rings. Both were found to have different impact on the origination of each ZDW.

There are requirements to locate the zero CD point at the λ of 1.55 μm or, in general, in the C-band [8]. Changing the air filling fraction and lattice pitch is not the only idea that could be used to optimize CD. For example elliptical holes instead of circular holes exhibit potential low and very stable CD properties, i.e. 0.6-1 ps/nm/km in the range from 1 μm to 1.9 μm [3].

Finally, there are a few examples that are worth to be noticed: a DCF in Ref. [27], with CD parameter being −1350 ps/nm/km at 1.55 μm; another one in Ref. [28] has CD of −440 to −480 ps/nm/km at the band of 1.5-1.62 μm [3].

3. STATE OF THE ART

The employment of the considered fibres is mainly in high-speed transmission systems using wavelength division multiplexing and signal recovery, offering transmission rates of even more than 1 Tb/s, where it is proved that non-optimized CD could destroy the pulse spreading, but as well it could generate some nonlinear effects, such as four wave mixing, among others. Although the properties of the fibres significantly differ from one another, the techniques used for optimized CD could be systematized so that one could design a fibre with desired CD value and slope at a specific λ.

Understanding of the mechanisms governing CD tailoring is necessary not only permits for the fibre design, CD suppression and avoidance, but also to predict the potential manufacturing tolerances. There are a few techniques used to obtain the expected CD and we describe them in more details in the following sections.

3.1. Fibre designs [1]

One of the techniques applied is to dope the fibre's core. By inserting small doped inclusion in a MOF's core, the index contrast with result in much larger mode confinement. Because the waveguide dispersion is related to it and because the length of propagated light waves is of the range that is comparable to the core size, the inserted region will be responsible for very low waveguide dispersion component.

There is another technique that could be used to optimize dispersion, it assumes to inject some liquids into the microstructured holes. In Ref. [29], the 4th ring of holes is doped by using liquids for this purpose. But liquid's properties are strongly dependent on temperature, which means that constant temperature must be ensured. This means that such a fibre could for example be used for some experiments in laboratory conditions (e.g. sensors), but they can't be used in outdoor installations of telecommunication systems [1].

Germanium dioxide is very often used to raise the effective index. In Ref. [13] and [30], the idea how to use it to dope the microstructured fibre's core is described in details. But this solution is not perfect in terms of effective area of the mode which is rather small in the doped fibres, as well as from the perspective of increased attenuation caused by material absorption, which is high for germanium dioxide [1].

Another technique to be described and systematized is the one using air holes located close to the fibre centre. The 1st mode could be confined within the core by the index contrast, and the effective diameter of a mode is then usually small enough to control CD. Stronger confinement could be obtained in a MOF using other type of a lattice, like for example octagonal, decagonal or spiral.

An alternative way, how to control the effective index, is to have core defects. A "defect core" is usually understood as one allowing light coupling to some other regions by some imperfection that is introduced intentionally. This technique uses weak interactions between the core-guided mode and a mode localized in an intentionally introduced defect of the crystal.

Finally, the idea of concentric cores could be considered. What feature in dual core technique is responsible for low CD? It is large value of n_{eff} produced by significant asymmetry between the two cores where light is coupled. Light is propagated in both cores concurrently. Dual cores are responsible for the phase matching at a certain λ, and consequently at this wavelength large CD parameter is obtained.

Low value of CD parameter is possible in concentric cores. They could be obtained by removing [7],[10] or reducing [7],[31] the air fraction in some rings of holes [1].

There are many aspects that could influence resultant CD; it could be coupling modes and the phase matching wavelength, at which CD control is possible, could be obtained in many ways, for example by fibre bending. We found these conclusions as interesting, because many works assume that the most important consequence of bending a fibre is increased attenuation. It could be shown that large negative value of CD parameter could be the accompanying process.

3.2. Extension of transmission band towards infrared — Fluoride fibres [32]

Dealing with the structural parameters of MOFs is a solution to find optimal CD. They could be enhanced by very careful optimizing the material composition. For example fluorides could be used as additives, but they could also be used as a background material instead of silicon dioxide, which could be evaluated as significant progress in the field [32]. The fluorides have low effective index, broad range of λs and low attenuation. One could consider the use of BaF_2, CaF_2 and more advanced ZBGA (ZrF_4-BaF_2-GdF_3-AlF_3) or ZBLAN (ZrF_4-BaF_2-LaF_3-AlF_3-NaF) [32]. In addition, because those materials are composite, it is possible to change their composition in a compound in order to obtain

n_{ff} value or its slope in the investigated area of λ. Last but not least, these materials are transparent in infrared region (as well as in the telecommunication bands) which makes them interesting for some applications in sensors. The properties of fluorides allow using them in MOFs [33]. They would lower the effective index of the cladding or the index at the second, larger concentric core. It is usually combined with doping the main core by using germania. This idea could be considered as a significant progress in fibre optic technology. Then a fibre with a W-type profile or refractive index could be created. For example, the core doped by using germanium dioxide would exhibit the n_{eff} of about 1.48 and concurrently the cladding index doped by using F-SiO_2 would exhibit the index of about 1.43. Similar solution (in general modified W-type index profile fibre) could be employed in systems with WDM [32],[34].

Material and structural parameters are key features in order to obtain exact CD at each λ, but the accuracy and quality of production could become a critical issue, especially in submicron fibres whose small holes with the diameter being less than one micron is problematic for some manufacturing technologies. All this combined with the application of new materials that should be processed in different way compared to siliceous materials could result in both attenuation and CD properties far from the expectations [32].

At the telecommunication band both ZBLAN and ZBGA showed similar attenuation properties like the one of silicon dioxide, but their CD properties significantly differ [35], in addition fluorides have different mechanical properties and they have to be fused at different temperature than temperature for silica. The theoretical attenuation could even be of 0.01 dB/km, in the range from 0.3 μm to 4.3 μm. The n_{eff} in telecom band is 1.47 – 1.52 [32, 36].

4. DESIGN APPROACHES

To describe propagation of light in a given optical glass, we could consider the constant of propagation of a wave that is related to the phase variation [32]:

$$k = (\beta - j\alpha) \tag{1}$$

The real part β and the imaginary part ω could be specified [37]:

$$\beta = \omega\sqrt{\frac{\mu\varepsilon}{2}\left[\sqrt{1+\left(\frac{\sigma}{\omega\varepsilon}\right)^2}+1\right]} \tag{2}$$

$$\alpha = \omega\sqrt{\frac{\mu\varepsilon}{2}\left[\sqrt{1+\left(\frac{\sigma}{\omega\varepsilon}\right)^2}-1\right]} \tag{3}$$

Where ω is angular velocity, σ is conductivity, a μ is permeability and ε is permittivity. Because MOFs are dielectric it could be assumed that:

$$\left(\frac{\sigma}{\omega\varepsilon}\right) << 1 \tag{4}$$

Both real and imaginary part of propagation constant in (3) could be simplified to a form:

$$\beta = \omega\sqrt{\mu\varepsilon} \tag{5}$$

$$\alpha = 1 \tag{6}$$

MOFs could be described by an effective index n_{eff}, which expresses "the refractive index at the boundary of two entities", which in this particular case are the core and the cladding [32]:

$$n_{eff} = \frac{\beta}{k_0} \tag{7}$$

β is real part of the constant of propagation of core, k_0 is the constant of propagation in vacuum. Propagation constant depends on λ, it could be written that:

$$k = \frac{2\pi}{\lambda} \tag{8}$$

Group velocity that integrates optical waves in an "envelope" could be considered to describe its velocity. It refers to CD and is less than the speed of electromagnetic wave in vacuum [32]:

$$v_g = \frac{\partial \omega}{\partial k} = \frac{c}{\sqrt{1 + \frac{\omega^2 \lambda^2}{4\pi^2 c^2}}} \tag{9}$$

Then group delay τ_g could be calculated as time necessary for the propagation of wave to certain distance with velocity of v_g (9):

$$\tau_g = \frac{1}{v_g} \tag{10}$$

In general, chromatic dispersion *CD* is known as the dependence of group delay τ_g on the wavelength λ, at which the signal is transmitted [38]:

$$CD(\lambda) = \frac{\partial \tau_g(\lambda)}{\partial(\lambda)} \tag{11}$$

In simulations it is more suitable to use material dispersion equation described by using Sellmeier approximation. CD equation suitable for Sellmeier approximations is following [39]:

$$CD(\lambda) = -\frac{\lambda}{c} \frac{\partial^2 Re\left[n_{eff}\right]}{\partial \lambda^2} \tag{12}$$

Designs of MOFs could be calculated by Finite-Difference Frequency Domain (FDFD) method. The distribution of light could be calculated by discretizing electric (13) and magnetic (14) field, it is described in Refs. [32, 40], and [41]:

$$\left(\nabla_t^2 + k_0^2\varepsilon_r\right)\boldsymbol{E}_t + \nabla_t\left(\varepsilon_r^{-1}\nabla_r\varepsilon_r \cdot \boldsymbol{E}_t\right) = \beta^2\boldsymbol{E}_t \tag{13}$$

$$\left(\nabla_t^2 + k_0^2\varepsilon_r\right)\boldsymbol{H}_t + \varepsilon_r^{-1}\nabla_t\varepsilon_r \times \left(\nabla_t \times \boldsymbol{H}_t\right) = \beta^2\boldsymbol{H}_t \tag{14}$$

k_0 is wavenumber in free space, β is real part of the propagation constant, ε_r is dielectric constant as in (5). The FDFD method uses the discretization scheme shown in Ref. [42][32].

FDFD is used in many commercial simulators of mode distribution and n_{eff} calculators. In most of them, it is possible to use user friendly graphical interface, but when a very accurate result is demanded, many iterations ought to be done by creating a loop using an appropriate scripting language. We use a mode solver from the Lumerical Inc. When the fibre's cross section is proposed and the simulation parameters and monitors are set, the simulation is run [3].

Creation of realistic models of sophisticated optical glasses is an interesting feature. To introduce materials to the simulator, we used the Sellmeier approximation [32, 43],[44]:

$$n^2(\lambda) = A + \sum_i \frac{B_i\lambda^2}{\lambda^2 - C_i^2} \tag{15}$$

where A, B_i, C_i are coefficients referring to index n. As a result, the wavelength evolution of refractive index could be plot by using the Sellmeier equation. The coefficients for the expanded version of Sellmeier equation (16) could found in [32][45]:

$$n^2 - 1 = \frac{B_1\lambda^2}{\left(\lambda^2 - C_2^2\right)} + \frac{B_3\lambda^2}{\left(\lambda^2 - C_4^2\right)} + \frac{B_5\lambda^2}{\left(\lambda^2 - C_6^2\right)} \quad (16)$$

where $B_{1,3,5}$ $C_{2,4,6}$ are material constants and λ is wavelength.

5. EXEMPLARY RESULTS

In this section, we present some selected fibre designs covering the wide area of telecom photonic fibres for CD tailoring, such as DCFs with large negative CD parameter, wideband fibres or slope compensating fibres, or fibre designs suitable for CD prevention, such as near zero CD fibres. We employ different design approaches and optimize them based on parametric investigation.

5.1. Dual core dispersion compensating fibre with large negative dispersion parameter [1]

A concentric core DCF based on microstructured optical fibre is proposed [1]. There is an assumption for the proposed structure to keep the 1st mode in a central core over the wide working spectrum of λ and in order to obtain low CD at 1.55 μm, and with low theoretical losses. The refractive index is optimized by making smaller the hole diameter in the 2nd ring. The design is specified in Table 1, the cross section is shown in Figure 1 [1].

Table 1. Parameters of a dual core MOF for CD suppression [1].

Quantity [unit]	Value	Value	Value	Value	Value
Lattice pitch Λ [μm]	1.55	1.52	1.50	1.48	1.45
Minimum D at 1.55 μm [ps/nm/km]	-1010	-1180	-1300	-1460	-1690
Core index [-]	1.44	1.44	1.44	1.44	1.44
Theoretical loss [dB/cm]	$9 \cdot 10^{-10}$	$7 \cdot 10^{-10}$	$3.8 \cdot 10^{-7}$	$3.8 \cdot 10^{-4}$	$7 \cdot 10^{-2}$
Full width at half maximum [nm]	142	132	129	125	122
Hole diameter d_1 [μm]	1.33	1.34	1.314	1.27	1.23
Hole diameter d_2 [μm]	0.55	0.54	0.574	0.63	0.67
Hole diameter d_3 [μm]	1.35	1.35	1.2	1.04	0.92
Number of rings [-]	6	6	6	6	6

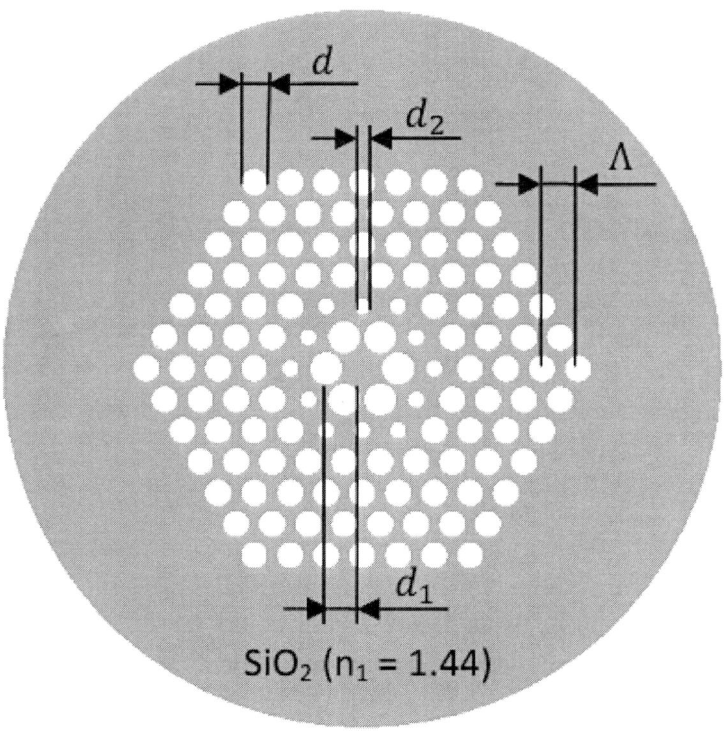

Figure 1. Structural parameters and the cross section of a dual core MOF for CD suppression purposes [1].

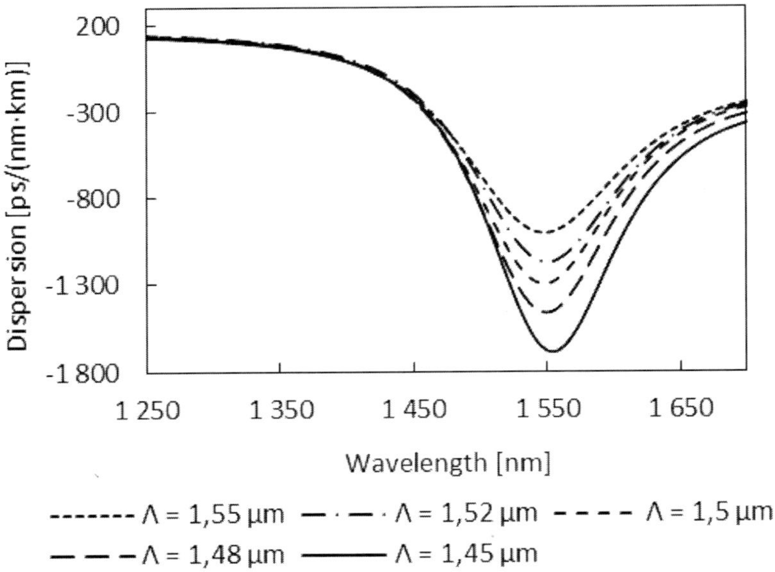

Figure 2. Chromatic dispersion in a dual core MOF for CD suppression at 1.55 µm [1].

CD of the considered fibre is shown in Figure 2. Obtained CD parameter was -1460 ps/nm/km. The designed fibre work in the C-band, and the lowest CD is at 1.55 µm. Theoretical losses are 3.8 10^{-4} dB/cm. It could be concerned as one of the advantages of the MOF, when we have a look at losses presented for some other fibres [1].

In order to obtain very accurate results, a number of iterations were performed to find the most suitable structural parameters of a fibre. One could conclude that smaller holes in the 3rd ring, and concurrently larger radius of holes in the 2nd ring could result in improved CD properties (Figure 3). Optimized solution could be found in Table 2 [1].

In Figure 3, it could be observed that λ at which there is minimum CD is function of normalized hole diameter d/Λ of the holes; CD is mostly sensitive to the holes located in the 2nd ring. To move the operating λ towards 1.5 µm, one has to increase the size of the holes in this ring.

Concurrently, the value of lowest CD is less. Summarizing, a one km long section of the designed DCF should be sufficient to tailor CD in a network that is created by using 75 km long conventional SMF with CD parameter being 17 ps/nm/km, (this value is in agreement with the recommendation of ITU-T for SMFs). The insertion losses of such a CD compensator are about 0.04 dB [1].

Table 2. Specification of parameters for dual core MOF for CD suppression [1].

Quantity [unit]	Value	Quantity [unit]	Value
Minimum CD [ps/nm/km]	-3290	Lattice pitch Λ [μm]	1.45
Core index [-]	1.44	Hole diam. d_1 [μm]	1.23
Theoretical losses [dB/cm]	7.50	Hole diam. d_2 [μm]	0.73
Number of rings [-]	6.00	Hole diam. d_3 [μm]	0.86

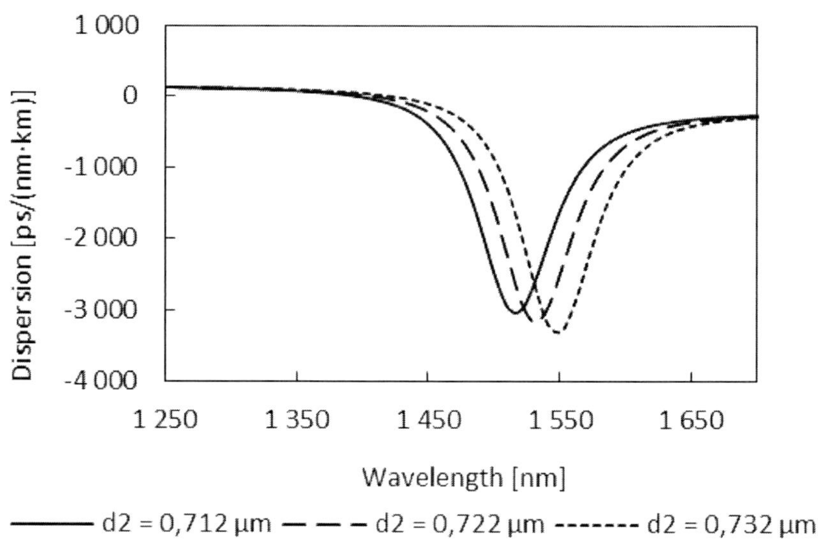

Figure 3. Dispersion in proposed dual core microstructured DCF [1].

5.2. DCF with optimized dispersion slope [3]

Another design is done with the scope on fibres that could compensate CD in each channel of a system using wavelength multiplexing. It means that the fibre should be wideband, and its CD mustn't be flat, but it should have exactly opposite CD to CD of a fibre that is used in a WDM system. The proposal is shown in Figure 4 and in Table 3 [3].

Table 3. Structural parameters of the HNPCF with dispersion evolution opposite to one in conventional MOF [3].

n_{eff} [-]	1.47	Hole diameter [μm]	3.64
Core index [-]	1.48	Normalized hole diameter d/Λ [-]	0.58
Core size [μm]	4.46	Number of rings [-]	5

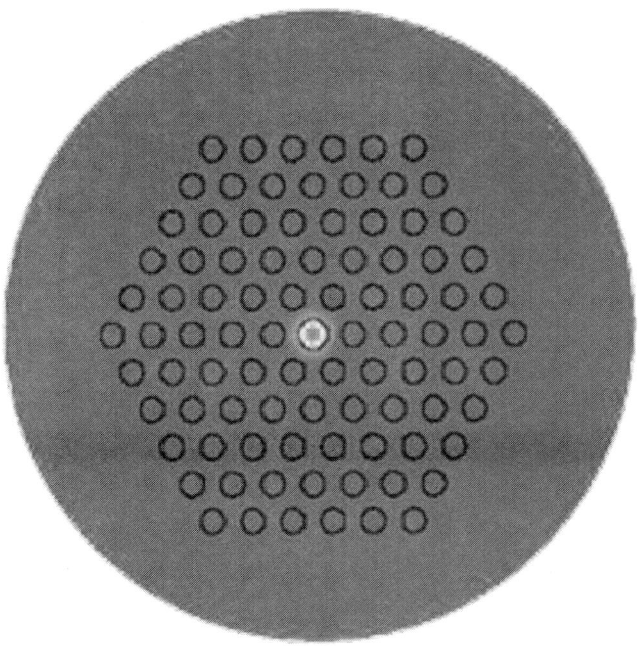

Figure 4. Cross section of a MOF with negative CD parameter [3].

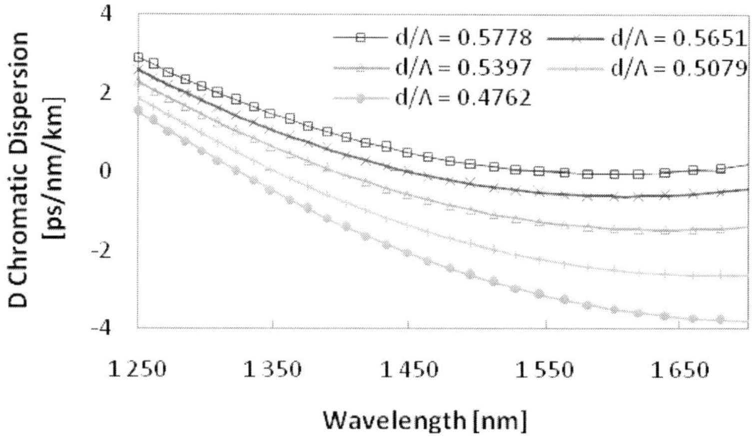

Figure 5. Reversed CD slope of the proposed DCF [3].

From Figure 5 it could be concluded that CD is reversed CD of conventional MOFs. The fibre is wideband. It could be optimized in terms of larger values of CD parameters, for example to exactly match CD evolution of conventional fibres. One shall also pay attention to the fact that larger d/Λ would result in multimode operation [3].

5.3. Dispersion flattened fibres and wideband dispersion compensating fibres [1][3]

A MOF with flat CD is demanded in design of transmission fibres (not suppression fibres) where it is d to have identical CD at each λ channel. Such a fibre could be doped in the core by using germania. The proposed core diameter could be 7.4 μm. In Figure 6, the 1st mode is kept in the core and the fibre is single-mode. Manufacturing of the MOF's core could be challenging, because of huge doping area; the fibre has large mode area. The proposed geometry is described in Table 4 [3].

Table 4. Structure parameters for HNPCF flattened CD curve [3].

Hole diameter d [μm]	1.32	Silica index [-]	1.46
Lattice pitch Λ [μm]	4.4	Propagating λ [μm]	1.55
Normalized hole diameter d/Λ [-]	0.3	Core size [μm]	7.4
Air index [-]	1	Effective cladding index at 1.55 μm	1.47
Core index [-]	1.48	Number of rings [-]	3

Obtained CD property could be evaluated as flat and oscillating around the value of -0.025 ps/nm/km. In Figure 7, comparison of CD in conventional MOF and the designed fibre is shown [3].

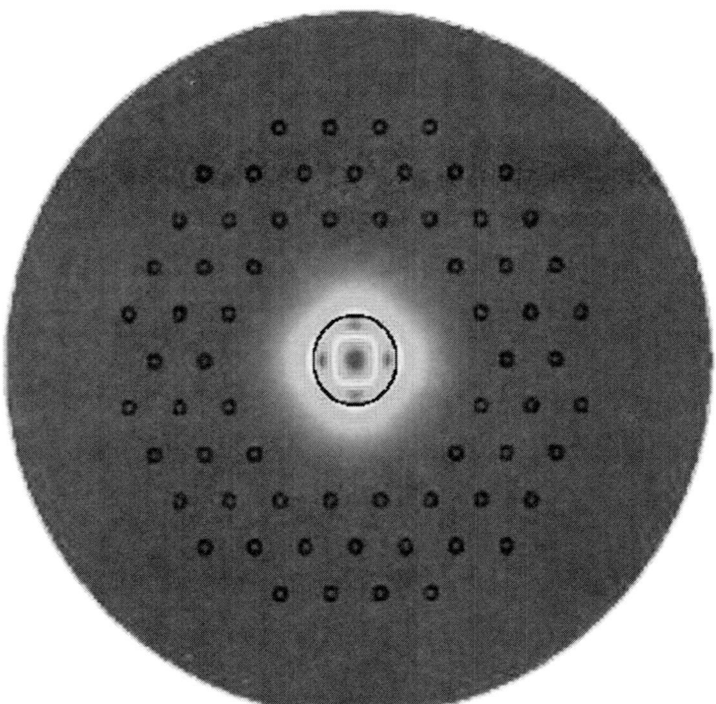

Figure 6. Wideband large mode area MOF with near zero flattened CD [3].

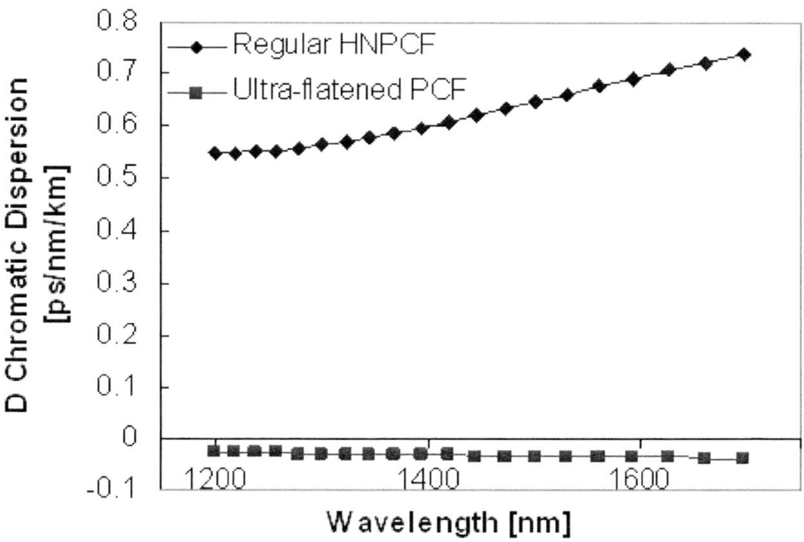

Figure 7. Dispersion in conventional microstructured fibre and proposed highly nonlinear MOF [3].

Wideband fibre could exhibit negative CD parameter, too. We propose a MOF for suppression in the band of 1.25-1.7 μm. We do not use any additives. Mode confinement is done by optimizing the geometry of the core and the normalized hole diameter d/Λ, the accepted view is that the core size should be small. In the 1st ring, there is $d_1/\Lambda=0.9$. The proposal could be found in Figure 8. Doping the core could additionally improve CD properties, but it would surely worsen attenuation. One should pay attention to the fact that optimization of CD shall be done in the context of attenuation properties. Optimizing one parameter and ignoring another is unacceptable in high-speed transmission system fibres [1].

Having larger holes in the 1st t ring, d_1, is responsible for lower CD. On the contrary, by making smaller all the other holes (d) results in increasing dispersion. In Table 5, optimized structure is shown and its cross section is in Figure 9 [1].

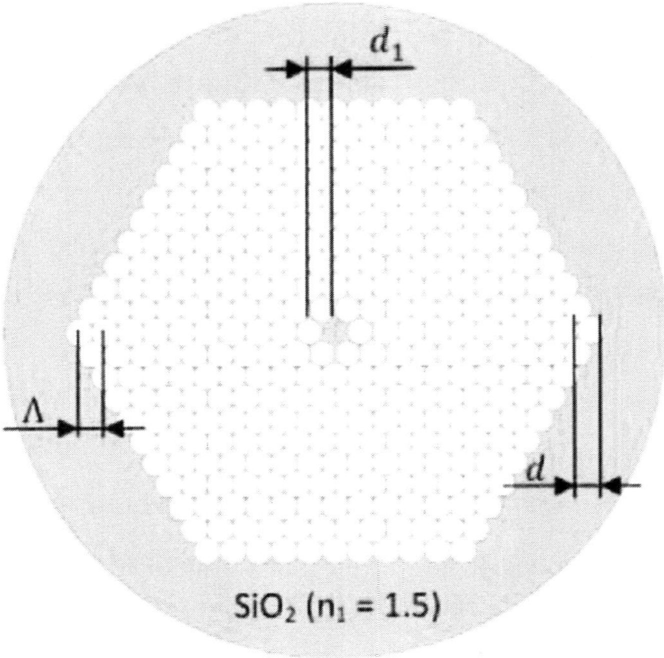

Figure 8. Cross section of designed microstructured fibre for wideband suppression of CD [1].

Table 5. Specification of parameters for wideband CD suppressing MOF [1].

Quantity [unit]	Value	Value	Value	Value	Value
CD [ps/nm/km]	-1580	-2040	-2259	-2094	-1930
Core index [-]	1.50	1.50	1.50	1.50	1.50
Theoretical losses [dB/cm]	$9.1 \cdot 10^{-6}$	$1.9 \cdot 10^{-3}$	$4.8 \cdot 10^{-2}$	$7 \cdot 10^{-2}$	$1.5 \cdot 10^{1}$
FWHM [nm]	Flat CD over 1.25-1.7 μm				
Lattice pitch [μm]	0.70	0.65	0.62	0.6	0.55
Hole diameter in 1st ring d_1 [μm]	0.70	0.65	0.62	0.6	0.55
Hole diameter d [μm]	0.70	0.65	0.62	0.6	0.55
Number of rings [-]	10	10	10	10	10

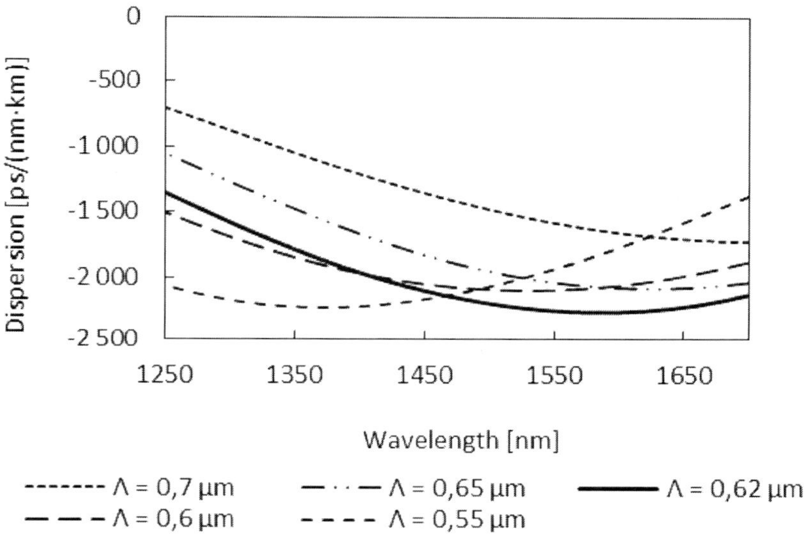

Figure 9. Resultant flattened negative value of CD over broad band of λs 1.25-1.7 μm (lattice pitch is a parameter) [1].

Concerning the values of CD, the best design is the one for Λ=0.62 μm, for which CD is -2259 ps/nm/km, obtained at 1.55 μm. Theoretical losses are 4.8 dB/cm. The optimized fibre is the one with Λ=0.70 μm, where CD parameter is -1580 ps/nm/km at 1.55 μm. Theoretical loss is lowered to 9.1 10-6 dB/cm. In this case it is possible to compensate dispersion of about 90 km of standard SMF.

5.4. Fluoride doped dispersion compensating fibres [32]

The considered fibre is a MOF employing the idea of a W-profile fibre with the core doped by using BaF_2 (n_{BaF2} = 1.468 to raise its refractive index) and containing three holes in the 1st ring doped by using CaF_2, n_{CaF2} = 1.426 to reduce effective cladding index. Increased index contrast is responsible for enhanced CD, as in eq. (12). As for example, models optical glasses could be expressed using coefficients shown in Table 6.

Table 6. Sellmeier coefficients for the fluoride additives used in investigated MOF [32].

Material	B_1	C_2	B_3	C_4	B_5	C_6
Silica	0.6961663	0.0684043	0.4079426	0.11624	0.897479	9.89616
BaF_2	0.6433560	0.0577890	0.5067620	0.109680	3.82610	46.3864
CaF_2	0.5675888	0.0502636	0.4710914	0.10039	3.848472	34.6490

The structural and material properties are in Table 7. Large radius of CaF_2 holes and the low core size are necessary for flat CD (Figure 10). CD is -413 ~ -415 ps/nm/km at 1.4 - 1.65 μm. Theoretical loss is 1.75 10^{-4} dB/cm. CaF_2 doped holes affect the CD (Figure 11). The larger are the doped holes, the lower is resultant CD [32].

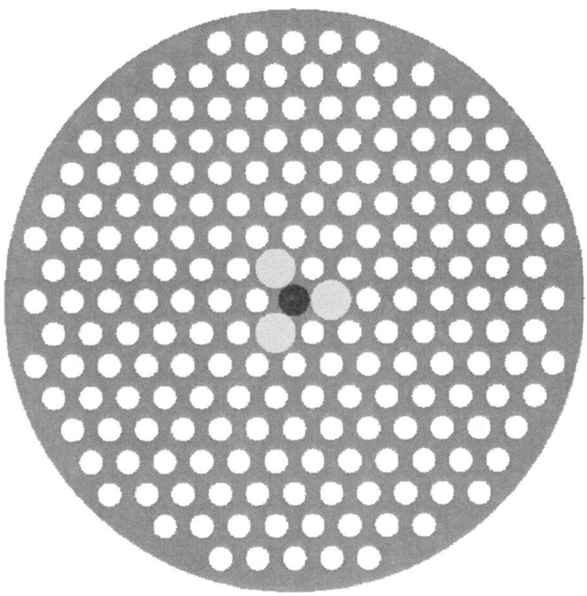

Figure 10. Fluoride-doped W-profile fibre's cross section for infrared and telecom band [32].

Table 7. Design specification of investigated W-profile MOF with doped holes in the 1st ring [32]

Proposed fibre:	
Air hole diameter d [μm]	0.4
Lattice pitch Λ [μm]	0.7
Core size [μm]	1.3
Doped hole diameter d_1 [μm]	0.76
Core dopant material	BaF_2 n = 1.468 at 1.55 μm
1st ring doping material	CaF_2 n = 1.426 at 1.55 μm
Normalized air-hole diameter d/Λ [-]	0.57

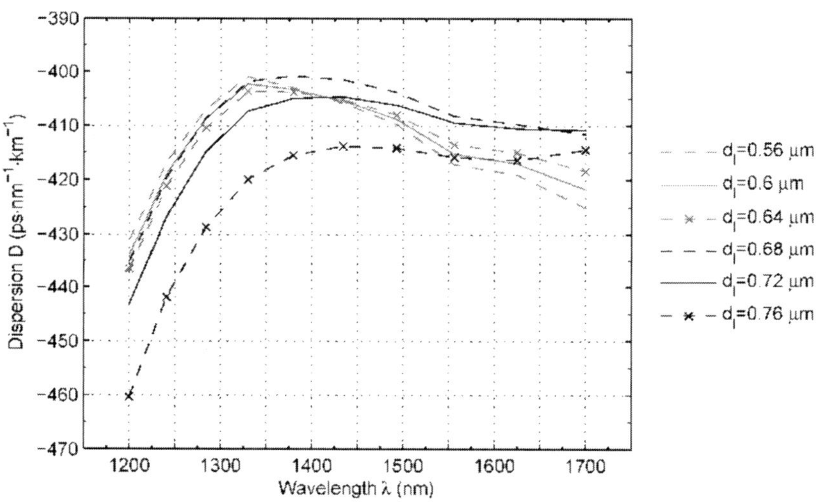

Figure 11. Wavelength evolution of CD with hole diameter of CaF_2 as a parameter [32].

The air holes influence the dispersion slope. Larger d/Λ would limit the confinement losses. An interesting feature is that we use fluorides, ZBGA or ZBLAN, not just to dope the cladding, but we propose to use it as a background material instead of silicon dioxide (Figure 12). As a result, optimization of CD is more flexible. Last but not least, temperature and mechanical properties of such a fibre wouldn't be worse [32].

Right combination and the composition of additives is responsible for optimized properties of a compound. Let us consider zirconium (Zr), barium (Ba), gadolinium (Gd) or aluminium (Al) that are the compounds of ZBGA material: ZrF_4 – BaF_2 – GdF_3 – AlF_3 [32].The n_{eff} of ZBGA material could be represented by modified Sellmeier equation (17) [46], using the coefficients in Table 8 [45].

$$n(\lambda)=\frac{A}{\lambda^4}+\frac{B}{\lambda^2}+C+D\lambda^2+E\lambda^4 \tag{17}$$

Table 8. Sellmeier coefficients for the fluoride-background ZBGA fibre [45].

Material	A	B	C	D	E
ZBGA at 25°C	2.98316e-6	3.39740e-3	6.81447e-3	-1.20276e-3	-5.48085e-6

ZBGA has larger n_{eff} comparing with silicon dioxide. Tailored CD is expected [32].

Parametric sweep performed for the hole diameter showed that flatten CD is possible, which is shown in Figure 6. For hole diameter being equal to 0.44 μm, the evolution of CD is similar to one obtained for silicon dioxide. Larger n_{eff} of the ZBGA fibre requires larger radius of air-holes in the cladding to tailor CD [32].

BaF_2 could be used in the core and CaF_2 in the 1st ring, ZBGA as a background material (Figure 13). CD parameter is then possible within the range of -435 ~ -438 ps/nm/km and over the entire telecommunication band. At the same time, attenuation properties are not worse than those obtainable in fibres with silica as a background material [32].

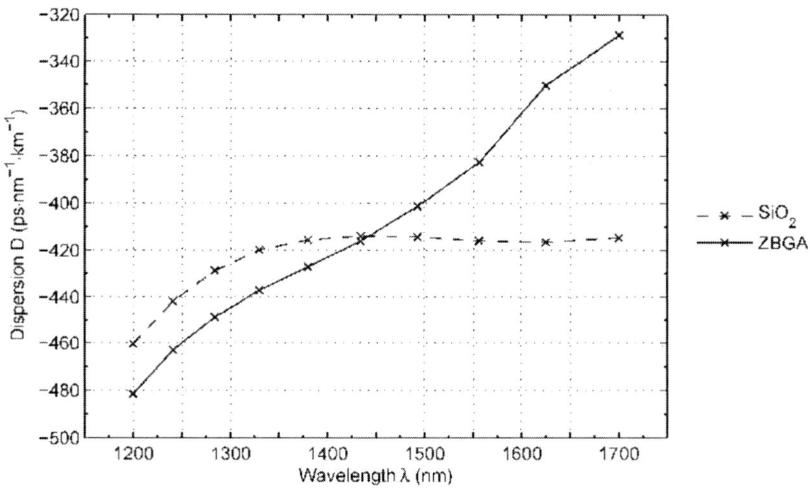

Figure 12. CD in SiO_2 and ZBGA used as a background material [32].

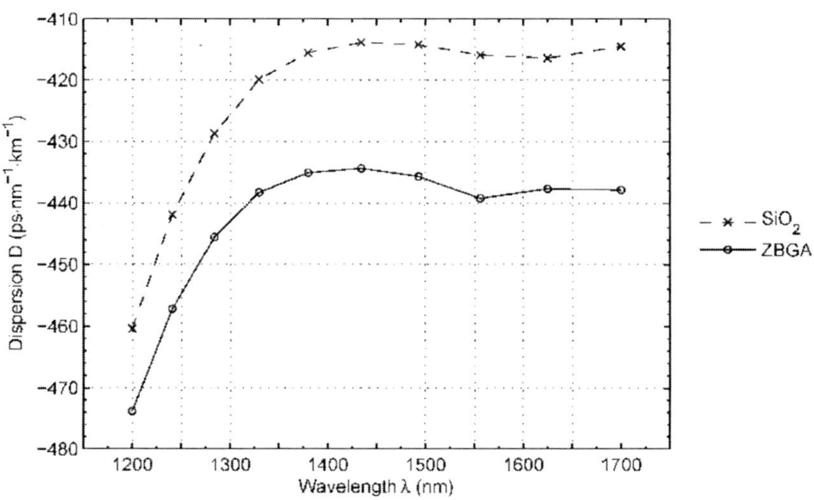

Figure 13. Proposed MOF with ZBGA material used as a background compared to a fibre with SiO_2 as a background [32].

5.5. Submicron dispersion compensating fibres with modified geometry [2]

One of the trends is to use a lattice with the pitch being less than one micrometre (so-called submicron lattice). Then, by careful adjustments of the diameter of holes (d_1-d_3 in Figure 14), where d_1 is hole diameter in the 1st ring of holes, one could obtain nearly zero CD for the λs from the window 1.25-1.7 μm. Hole diameter and lattice pitch are submicron. Then, high confinement of light waves and strong waveguide dispersion is possible. The geometrical parameters of the fibre are in Table 9 and in Figure 14 [2]. CD and loss are simulated for the 1st light mode. Confinement losses are 6.10^{-6} dB/km (Figure 15). Wavelength evolution of CD is shown in Figure 15 [2].

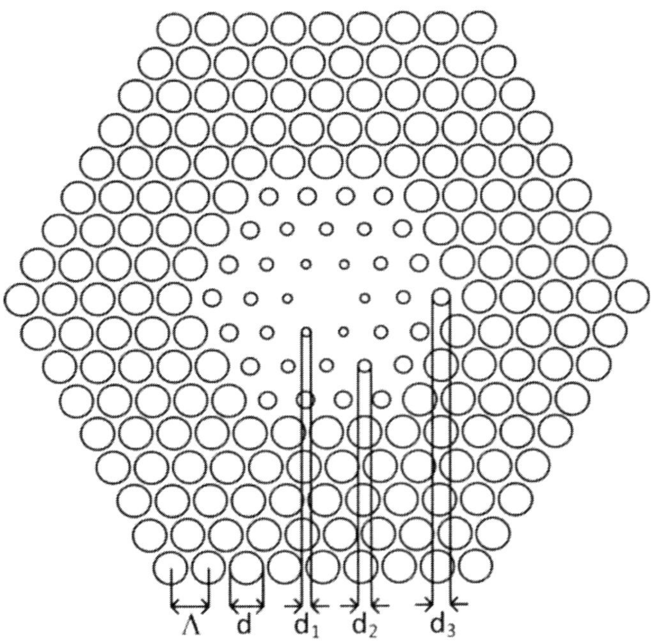

Figure 14. Cross section of a near zero CD flattened MOF with modified rings [2].

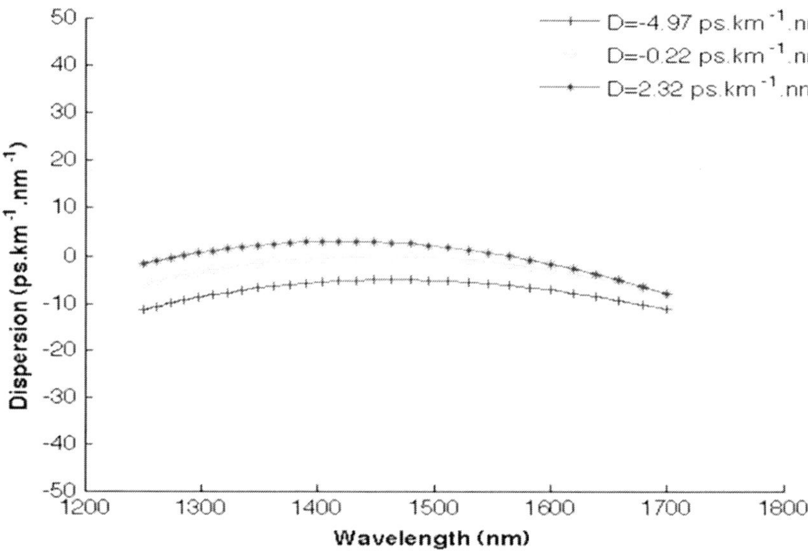

Figure 15. Near zero flattened CD for wideband utilization in telecommunications [2].

Obtained CD parameter is from -7.7 ps/nm/km to 3.1 ps/nm/km at the band of (1.25, 1.7 μm) and its average value is around 0.51 ps/nm/km. In the C-band, it is 1.35±0.46 ps/nm/km. For both C and L bands CD is 0.12±1.32ps/nm/km [2]. Results for parametric iterations for air fraction changed in three most internal rings are shown in Figures 15 and 16. From the results summarized in Table 10 it could be concluded that the fibre has CD slope of -0.09 ps.nm^{-2}.km^{-1} is obtained for little variation of d_1 from other holes and is suited to the slope of conventional ITU-T G.657 fibres. The slope is less than 0.09 ps.nm^{-2}.km^{-1} [2].

Table 9. Geometrical parameters of a fibre with nearly zero flattened CD [2].

d_1	d_2	d_3	d = const.	Λ = const.	CD at 1.55 μm
μm	μm	μm	μm	μm	ps/nm/km
0.134	0.305	0.318	0.355	0.8	2.32
0.134	0.298	0.325	0.355	0.8	-0.22
0.134	0.290	0.325	0.355	0.8	-4.97

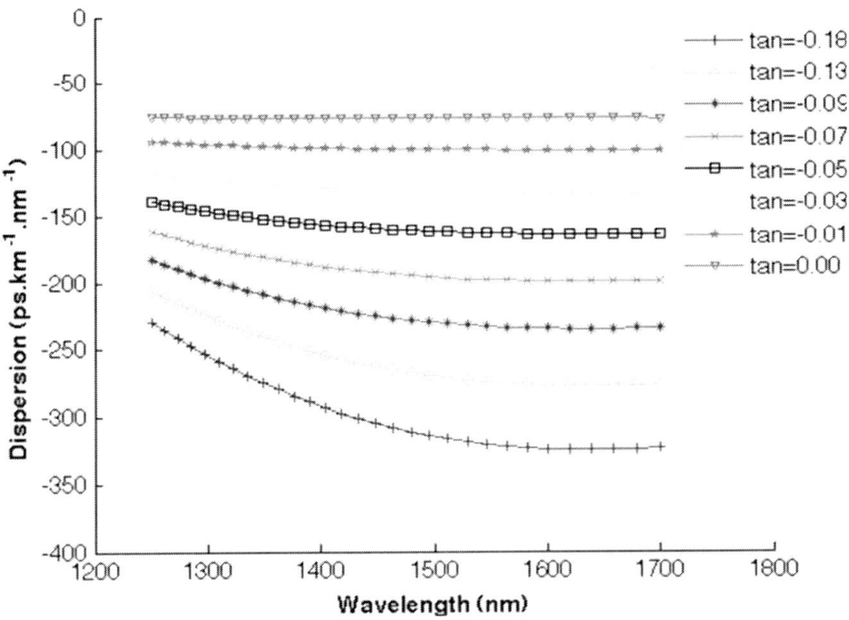

Figure 16. Optimized CD slope by varying the hole diameter in the 1st ring [2].

Larger CD slope is here obtained by adjusting the value of d_3 (see Table 11, Fig. 16). At d_1 being fixed, linear wavelength evolution of CD is possible. It becomes nonlinear when the slope exceeds -1 ps.nm^{-2}.km^{-1}. The obtained property could be applied in CD suppression performed on a fibre with the slope of 0.5 ps.nm^{-2}.km^{-1} [2].

Table 10. Optimized dispersion slope by varying diameter of holes in the 1st ring [2].

	d_1	d_2, constant	d_3, constant	d, constant	CD slope
	μm	μm	μm	μm	ps.nm^{-2}.km^{-1}
hole diameter of the 1st ring	0.18	0.31	0.355	0.355	0.01
	0.20	0.31	0.355	0.355	-0.03
	0.22	0.31	0.355	0.355	-0.07
	0.24	0.31	0.355	0.355	-0.13
	0.25	0.31	0.355	0.355	-0.18

Table 11. Large CD slope obtained by tuning the hole size in the 3rd ring [2].

	d_1 = const.	d_2 = d, constant	d_3	d, constant	CD slope
	μm	μm	μm	μm	ps.nm^{-2}.km^{-1}
hole diameter of the 3rd ring	0.265	0.355	0.35	0.355	-0.39
	0.265	0.355	0.27	0.355	-0.63
	0.265	0.355	0.23	0.355	-0.85
	0.265	0.355	0.17	0.355	-1.32

6. FINAL CONCLUSIONS

We described optical fibres from the perspective of their CD properties, which has to be solved in telecommunication. The mechanisms governing CD properties in telecom fibres are shown. The considered fibres are suitable for potential suppression of group velocity dispersion, utilized mainly in the C-band (dual core fibres with high negative CD parameter) and in wideband applications (CD flattened MOFs and exact slope suppression fibres).

A special family of fibres is so-called fluoride fibres that have CD optimized through material dispersion. This idea is combined with the proposal of a W-type fibre, where particular regions are doped by using different additives, including fluorides, among others. ZBLAN material offers broader range with low attenuation (telecom and infrared range).

It makes them very attractive for getting eventually applied in spectrometry or in applications in fibre-optic sensors.

The use of fluorides assessed their flexibility in CD optimization and potential use in wideband CD suppression. Last but not least, it has been shown that optimization of fibre's CD slope is possible in submicron lattice pitch without additives, potentially for exact slope suppression of standard International Telecommunication Union fibres.

ACKNOWLEDGEMENTS

This work has been supported by the CTU foundation, SGS13-201-OHK3-3T-13.

REFERENCES

1. Lucki, M., Zeleny, R. Broadband Dispersion Compensating Photonic Crystal Fibre: conference proceedings, SPIE Vol. 8306. Bellingham (Washington): SPIE, 83060Z-1-83060Z-6, 2011.

2. Lucki, M., Jiruse, D., Kraus, S. Single-Material Submicron Microstructured Fibres for Broadband Applications in Exact Slope Compensation or Zero- Dispersion Propagation: conference proceedings, 14th International Conference on Transparent Optical Networks (ICTON). Piscataway: IEEE, Tu.P.28., 2012.

3. Lucki, M. Optimization of Microstructured Fibre for Dispersion Compensation Purposes: conference proceedings, 2011 13th International Conference on Transparent Optical Networks (ICTON). Piscataway: IEEE, p. Tu.P.12, 2011.

4. Veng M., et al. Dispersion compensating fibres. Opt. Fibre Technol. 2000; 6 164-80.

5. Antos A., Smith D. Design and characterization of dispersion compensating fibre based on the LP01 mode. J. Lightwave Technol. 1994; 12(10) 1739-1745.

6. Ferrando A., et al. Nearly zero ultraflattened dispersion in photonic crystal fibres. Opt. Lett.2000; 25 790-792.

7. Ni Y. et al. Dual-Core Photonic Crystal Fibre for Dispersion Compensation. IEEE Photonics Technology Letters 2004; 16(6) 1516-1518.

8. Liu Z. et al. A broadband ultra flattened CD microstructured fibre for optical communications. Optics Communications 2007; 272(1) 92-96.

9. Fujisawa T. et al. Chromatic dispersion profile optimization of dual-concentric-core photonic crystal fibres for broadband dispersion compensation. Opt. Express 2006; 14(2) 893-900.

10. Yang S. et al. Theoretical study and experimental production of high negative dispersion photonic crystal fibre with large area mode field. Opt. Express 2006; 14(7) 3015-3023.

11. Nejad S. M., Ehtehsami N. A Novel Design to Compensate Dispersion for Square-lattice Photonic Crystal Fibre over E to L Wavelength Bands: conference proceedings, Communication Systems, Networks and Digital Signal Processing Symposium, 654-658, 2010.

12. Hosaka T. et al. Dispersion of pure GeO2 glass core and F-doped GeO2 glass cladding single-mode opticalfibre. Eelectron. Letters 1987; 23(1) 24-26.

13. Hoo Y. et al. Design of photonic crystal fibres with ultra-low, ultra-flattened chromatic dispersion. Optics Communications 2004; 242(4-6) 327-332.

14. Ono-Kuwahara M. et al. Fluorine-doped silica fibre with high transparency and resistivity to deep ultra violet light, Lasers and Electro-Optics Lasers and Electro-Optics: conference proceedings,

Conference on Quantum Electronics and Laser Science. CLEO/QELS 1-2, 2008.

15. Olyaee S., Taghipour F. Ultra-Flattened Dispersion Photonic Crystal Fibre with Low Confinement Loss: conference proceedings, Proceedings of the 11th International Conference on Telecommunications, pp. 531-534, 2011.

16. Peckham D. W. et al. Reduced Dispersion Slope, Non-Zero Dispersion Fibre: conference proceedings, European Conferencce and Exhibition on Optical Communication,139-140, 1998.

17. Aikawa K. et al. High-performance Dispersion-slope and Dispersion Compensation Modules. Fujikura Technical Review 2002; 31.

18. Andersen P. A., Paulsen H. N., and Larsen J. J. A photonic crystal fibre with zero dispersion at 1064 nm. Optical Communication 2002; 2 1-2.

19. Hansen K. P. Introduction to nonlinear photonic crystal fibres. J. Opt. Fibre Commun. 2005; 2 226-254.

20. Ferrando A., Silvestre E., and Andrés P. Designing the properties of dispersion flattened photonic crystal fibres. Opt. Express 2001; 9(13) 687-697.

21. Lægsgaard J., Asger Mortensen N., and Bjarklev A. Mode areas and field energy distribution in honeycomb photonic bandgap fibres. J. Opt. Soc. Am. B 2003; 20(10) 2037-2045.

22. Birks T. et al. Dispersion compensation using single-material fibres. IEEE Photonics Technology Letters 1999; 11 674-676.

23. Haxha S., Ademgil H. Novel design of photonic crystal fibres with low confinement losses, nearly zero ultra-flatted chromatic dispersion, negative chromatic dispersion and improved effective mode area. Optics Communications 2008; 15(2) 278-286.

24. Saitoh K. et al. Chromatic dispersion control in photonic crystal fibres: application to ultra-flattened dispersion. Opt. Express 2003; 11(8).

25. Razzak S. M. A. Proposal for Highly Nonlinear Dispersion Flattened Octagonal Photonic Crystal Fibres. IEEE Photonics Technology Letters 2010; 20(4).

26. Razzak S. M. A. et al. Chromatic Dispersion Properties of a Decagonal Photonic Crystal Fibre: conference proceedings, International Conference on Information and Communications Technology, 159-162, 2007.

27. Ming W., et al. Broadband dispersion compensating fibre using index-guiding photonic crystal fibre with defected core. Chin. Opt. Lett 2008; 6 22-24.

28. Zsigri B., Lægsgaard J., Bjarklev A. A novel photonic crystal fibre design for dispersion compensation. J. Opt. A: Pure Appl. Opt. 2004; 6 717-720.

29. Yu Ch. et al. Tunable dual-core liquid-filled photonic crystal fibres for dispersion compensation. Optics Express 4443 2008; 16(17) 4443-4451.

30. Zhong, Q., Inniss, D. Characterisation of lightguiding structure of optical fibres by atomic force microscopy. J. Lightwave. Technol. 1994; 12 1517-1523.

31. Huttunen, A., Torma, P. Optimization of dual-core and microstructure fibre geometries for dispersion compensation and large mode area. Optics Express 2005; 13(2) 627-635.

32. Lucki, M., Zeleny, R.: Broadband submicron flattened dispersion compensating fibre with asymmetrical fluoride doped core: conference proceedings, Micro-structured and Specialty Optical Fibres II. Bellingham: SPIE, 87750M-1-87750M-8, 2013.

33. Tefelska, M. et al. Propagation Effects in Photonic Liquid Crystal Fibres with a Complex Structure. Optical and Acoustical Methods in Science and Technology, 2010; 118(6) 1259-1261.

34. Petruzzi, P., Lowry, C., Goldhar, J., Sivanesan, P., Dispersion compensation using only Fibre Bragg Gratings: conference proceedings, Optical Fibre Communication Conference 4, 14-16, 1999.

35. Saad, M. Fluoride glass fibres. Photonics Society Summer Topical Meeting Series 81-82; 2011.

36. Tran, D., Sigel, G., Bendow, B. Heavy metal fluoride glasses and fibres: A review. Journal of Lighwave Technology 1984; 2 566-586.

37. Poli, F., Cucinotta, A., Selleri, S. Photonic Crystal Fibres: Properties and Applications. Springer Series in Materials Science 102. The Nederlands; 2007.

38. Hájek, M., Holomeček, P. Chromatická disperze jednovidových optických vláken a její měření Mikrokom; 2006.

39. Saitoh, K., Koshiba, M. Chromatic dispersion control in photonic crystal fibres: application to ultra-flattened dispersion. Optics Express 2003; 11(8) 843-852.

40. Stern, M. S. Semivectorial polarised finite difference method for optical waveguides with arbitary index profiles. Optoelectronics, IEE Proceedings 1988; 135 56-63.

41. Huang, W. P., Xu, C. L. Simulation of three-dimensional optical waveguides by a full-vector beam propagation method. IEEE Journal of Quantum Electronics 1993; 29 2639-2649.

42. Yee, K. Numerical solution of initial boundary value problems involving maxwell's equations in isotropic media. IEEE Transactions on Antennas and Propagation 1966; 14 302-307.

43. Ghosh, G., Yajima, H. Pressure-dependent Sellmeier coefficients and material dispersions for silica fibre glass. Journal of Lightwave Technology 1998; 16 2002-2005.

44. Iwasaki, T., Endo, M., Ghosh, G. Temperature-dependent Sellmeier coefficients and chromatic dispersions for some optical fibre glasses. Journal of Lightwave Technology 1994; 12 1338-1342.

45. Weber. J. M. et al. Handbook of Optical Materials. The CRC Press Laser and Optical Science and Technology Series; 2003.

46. Mitachi, S. Dispersion measurement on fluoride glasses and fibres. Lightwave Technology Journal 1989; 7(8) 1256-1263.

CHAPTER 3

Advances on Optical Fiber Sensors

Luciano Mescia and Francesco Prudenzano *

Dipartimento di Ingegneria Elettrica e dell'Informazione,
Politecnico di Bari Via E. Orabona 4, 70125 Bari, Italy

ABSTRACT

In this review paper some recent advances on optical fiber sensors are reported. In particular, fiber Bragg grating (FBG), long period gratings (LPGs), evanescent field and hollow core optical fiber sensors are mentioned. Examples of recent optical fiber sensors for the measurement of strain, temperature, displacement, air flow, pressure, liquid-level, magnetic field, and the determination of methadone, hydrocarbons, ethanol, and sucrose are briefly described.

KEYWORDS

laser; optical fiber; microstructured optical fibers

1. INTRODUCTION

Almost four decades have passed since the research on optical fiber sensors began in earnest. Various approaches and technologies have been utilized for measuring a number of different physical parameters, but only some types of optical fiber sensors are commercially interesting. In fact, in many cases, optical fiber sensor systems are not available in a complete form, i.e., including both detecting and signal-processing electronics. However, their future is very promising because they exhibit well known advantages such as compactness, immunity to electromagnetic interference and to ionizing radiation (γ-ray, X-ray etc.), high sensitivity, large bandwidth, and minimum weight. These properties, make optical fiber sensors key photonic devices in radiative environments, like nuclear power plants, where the detection and evaluation of radiation levels and temperature changes are very important, especially in case of accidental constraints [1]. Optical fiber sensors have been developed to measure strain, temperature, pressure, current, voltage, gas, chemical contaminant, rotation, vibration, acceleration, bending, torsion, displacement, and biomolecules [2].

Chemical optical fiber sensors for the detection of polluting substances in water, soil, and air, such as the biological sensors, optimized for medicine applications, have attracted noticeable interest because they are directly related to the quality of human life. They have been investigated as key elements of environmental monitoring systems, with the aim of preventing environmental catastrophes, and also as novel techniques for medical diagnosis. In particular, optical fiber sensor systems can avoid the collection of samples to be investigated in appositely equipped laboratories. They provide means of overcoming the drawbacks exhibited by ex-situ techniques which are invasive, time consuming and expensive. In fact, optical fiber chemical/biological sensors, enable pollution detection and/or medical diagnostic monitoring via an in-situ and slightly invasive technique.

Optical spectroscopy of water, atmosphere, soil or biological samples, is commonly performed in chemical laboratories and is based on Laser

Induced Fluorescence (LIF) or Raman effects [3]. LIF or Raman equipment is generally expensive due to both the utilized light sources (and related optical components) and to the sophisticated processing electronics. In order to develop low-cost pollution and bio-medical monitoring, a promising approach concerns the optimization of microstructured optical fibers (MOFs) for sensing. In fact, in recent years, a number of sensors made of suitable MOFs have been theoretically and experimentally investigated for the measurement of a large variety of physical and chemical parameters [4,5,6,7,8,9,10,11,12].

Both conventional fiber optic and MOF sensors can be based on fiber gratings, interferometers, scattering/reflecting, Faraday rotation, fiber-optic gyroscopes, fluorescence, luminescence, and interaction of evanescent electromagnetic field structures.

Most of the aforesaid fiber sensors employ optical gratings which, under phase matching conditions, can couple different propagation modes. Optical fiber gratings can be fabricated by employing different techniques, e.g., the exposure of the optical fiber core to an intense optical interference pattern [13]. Permanent gratings in an optical fiber were first obtained by Hill et al. [14]. In [14] an Argon-ion laser radiation, at 488 nm, was launched into a germania-doped fiber; after some minutes most of the launched light was reflected. Successively, strain and temperature tuning of the fiber grating were investigated and suitable spectral measurements confirmed the fabrication of a fiber Bragg grating (FBG) via the photosensitivity effect. In fact, a coherent high power laser beam can interfere with the light reflected back from a discontinuity to produce a standing wave pattern within the optical fiber. If the optical fiber, after ultraviolet light exposure, is suitably annealed the grating becomes permanent. It was demonstrated that the gratings facilitated coupling to higher-order leaky modes. After this experiment, numerous scientific papers on FBGs and their applications were produced.

Different techniques have been utilized for grating writing. In the holographic technique, two overlapping ultraviolet light beams interfere producing a periodic interference pattern on the side of the fibers [13].

Therefore a periodic refractive index grating is written in the core because the fiber cladding is transparent to the ultraviolet light whereas the fiber core, generally germanium doped, highly absorbs the ultraviolet light. The phase mask technique is based on the fabrication of a flat slab of silica glass, transparent to ultraviolet light but having one of the flat surfaces with a one dimensional periodic corrugation. This corrugation, can be obtained via an etching technique and by using photolithographic techniques. The optical fiber is placed close to the corrugations of the phase mask and then irradiated by ultraviolet light which is incident normal to the phase mask [13].

FBGs are now commercially available and they find extensive application in many areas such as, filtering, routing, lasing and amplification of optical signals, and sensing of various measurands.FBG-based wavelength interrogation modules and broadband light sources, have been optimized for the safety monitoring of large structures, e.g., bridges, towers, dams, and railways [15,16,17,18,19,20,21].

Long period gratings (LPGs) have been exploited in the construction of filters, gain flatteners for erbium-doped fiber amplifiers, dispersion compensators, sensors, etc. [22,23,24,25,26,27, 28,29,30,31,32,33, 34, 35,36,37]. They have been fabricated in MOF, by employing different techniques, to obtain strain or refractive index sensors [28,29]. Moreover, LPGs were fabricated in all-solid photonic bandgap fibers (PBGFs) by UV illumination [36]. In particular, the exploitation of a tunable slit with a high-precision motorized linear stage and a suitable PBGF laser pulse exposition allowed an optimized LPG fabrication process [36]. The first fabrication of femtosecond-laser-drilled structural LPGs in a large mode area PCF was reported in [37]. Different techniques allowing grating inscription into the MOF core and cladding have been demonstrated in a number of experimental works [23,25,26,30,35,36,37].

In this paper, some recent advances on optical fiber sensors are reported. Specifically, only a few recent investigations are cited for FBGs, LPGs and evanescent field sensors.

2. OPTICAL FIBER GRATINGS AS KEY ELEMENTS OF SENSORS

The core refractive index perturbation $\Delta n(z)$ along the fiber longitudinal direction z, constituting a grating, can be expressed as reported in [31]:

$$\Delta n(z) = n_1(z) - n_1 = n_1\sigma(z)\left[1 + m\sin\left(\frac{2\pi z}{\Lambda}\right)\right] \quad (1)$$

where n_1 is the core refractive index, Λ is the grating period, m is the induced-index fringe modulation, and $\sigma(z)$ is the slowly varying envelope of the grating.

The well-known coupled mode theory CMT can be employed to find the equations describing the electromagnetic field propagation along the grating, by considering the slowly varying envelope approximation (SVEA) [31]. Coupled equations can be correctly written when grating inscription induces a weak perturbation on the propagation modes and the electromagnetic field amplitudes of the forward and backward modes can be calculated.

A strong mode interaction can be obtained in the grating at the wavelength where both mode phase matching (phase synchronism) and a sufficient mode overlap occur. Therefore, different modes can exchange power between them. The local reflectivity is the complex ratio of the forward and backward wave amplitudes [13]. FBGs, characterized by a short period, typically couple the forward propagating core mode to the backward propagating one. Chirped fiber gratings exhibit a large reflection spectrum. They are generally utilized when the phenomenon that each wavelength component is reflected at different positions in different delay times must be exploited for the fabrication of specific devices [2]. Tilted fiber gratings exhibit different characteristics and they can be used to couple the forward propagating core mode to the backward one and to a backward propagating cladding mode. LPGs, characterized by a long period, are generally designed to allow the power exchange of a forward propagating core mode with the power of selected forward propagating cladding modes.

When a core mode is coupled with a cladding mode the transferred electromagnetic energy is affected by the mode phase mismatch (displacement from the phase matching condition) which can be due to a variation of a number of different measurands such as wavelength, temperature, stress, fiber bending etc. Therefore optical sensors can be obtained.

2.1. Fiber Bragg Grating Sensors

An FBG is a distributed mirror at the Bragg wavelength, i.e., at the phase-matching condition $\lambda_B = 2n_{eff}/\Lambda$, where λ_B is the Bragg wavelength, n_{eff} is the effective index of the interfering guided mode [16]. Fiber laser cavity can be typically obtained by employing two different mirror implementations: (i) distributed Bragg reflector (DBR) formed by two cascaded wavelength-matched FBGs; and (ii) distributed feedback Bragg (DFB) structure where a phase-shifted FBG is distributed in an active fiber [17,18,19]. DBR and DFB laser sensors attract considerable interest because of their excellent characteristics [20,21]. Moreover, fiber grating laser sensors can be based on wavelength encoding or on a polarimetric heterodyning sensor.

FBGs allow fine control of fiber laser cavity by finely tuning their distributed reflectivity. This potential is strategic in the optimal design of both conventional and MOF fiber lasers [38,39,40,41,42,43,44,45] and therefore in the design of DBR and DFB laser sensors.

A heavy limitation of FBG strain sensors lies in their high temperature cross-sensitivity, this drawback is worsened by the difficulty to measure strain and temperature simultaneously. In [46] a hybrid sensor overcoming the aforesaid limitation has been proposed. It operates in the intensity domain, by converting into intensity changes two different items: (i) the polarization information from a polarization-maintaining photonic crystal fiber PM-PCF fiber sensor; and (ii) the wavelength information from FBG. A 3 dB coupler has been used to split the signals between the FBG sensor and the polarimetric sensor. A polarizer and an analyzer at the input and output of the PM-PCF have been used to measure the polarization variation due to the strain and temperature. A macrobend fiber-based edge filter has been used to convert the FBG wavelength variations due to the strain and temperature into light intensity changes. The output powers were measured by employing two photodiodes [46]. Also in displacement measurements, the FBG sensor output can be obtained by using an edge filter to interrogate the resonant wavelength [47] or a cantilever to transfer/convert the displacement into the chirp profile of the FBG [48].

Optical fiber sensors based on cladding-mode backward recoupling constitute an interesting technique to obtain high sensitivities [49,50,51,52,53]. In fact, an optical fiber cladding mode shows much higher sensitivity to micro deformations, like structural bending, compared to a fundamental core mode. For example, in [54] a structure for cladding-mode backward recoupling has been fabricated by including the up-taper and the Bragg grating as shown in Figure 1. The up-taper microscope photo is reported in Figure 2. Cladding-mode backward recoupling is the key principle on which the sensor is based. The structure is illuminated by a broad-band source from the same interrogation fiber. The up taper is designed to allow mode coupling between copropagating modes. The Bragg grating allows wavelength-selective mode coupling between counter-propagating modes. At the up-taper, the core mode exhibits loss and the cladding modes are excited. The excited cladding modes travel to the fiber grating with low attenuation if the sensor is kept straight. As described in [54], the guided core mode is coupled to the backward-propagating cladding ones by the grating. Then the cladding modes are recoupled to the guided core mode at the up taper. The cladding modes excited at the up taper are recoupled to the guided core mode by the fiber grating. Following this, the guided core mode travels into the interrogation fiber.

Mode coupling between counter-propagating cladding modes can be neglected. Figure 3 illustrates cladding mode power and core mode power versus displacement and cladding mode power versus temperature. An improvement of the sensor performance is obtained by employing suitable uniform chirped FBGs (CFBGs) as illustrated in Figure 4. With the same measurement set-up, by employing the CFBG the maximum reflection power is increased from 800 nW (for the FBG) to 176 μW (for the CFBG).

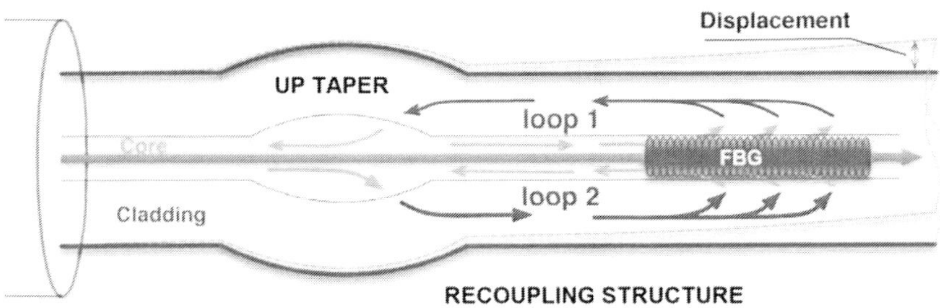

Figure 1. Schematic of cladding-mode backward recoupling structure including the up taper and the fiber Bragg grating proposed in [54].

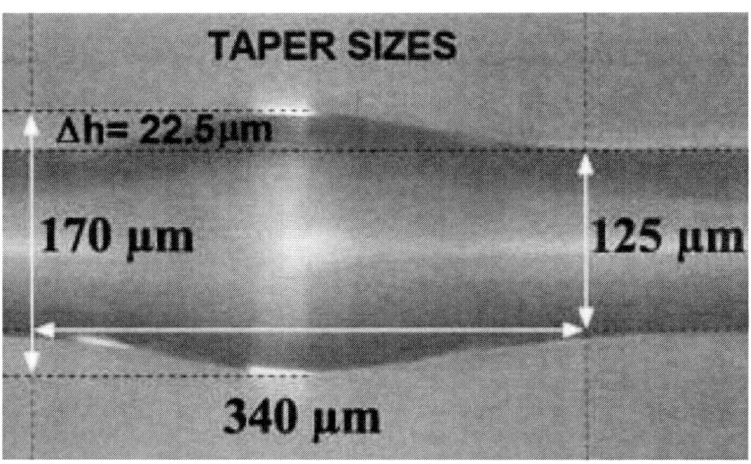

Figure 2. Microscope photo of the up taper constructed in [54].

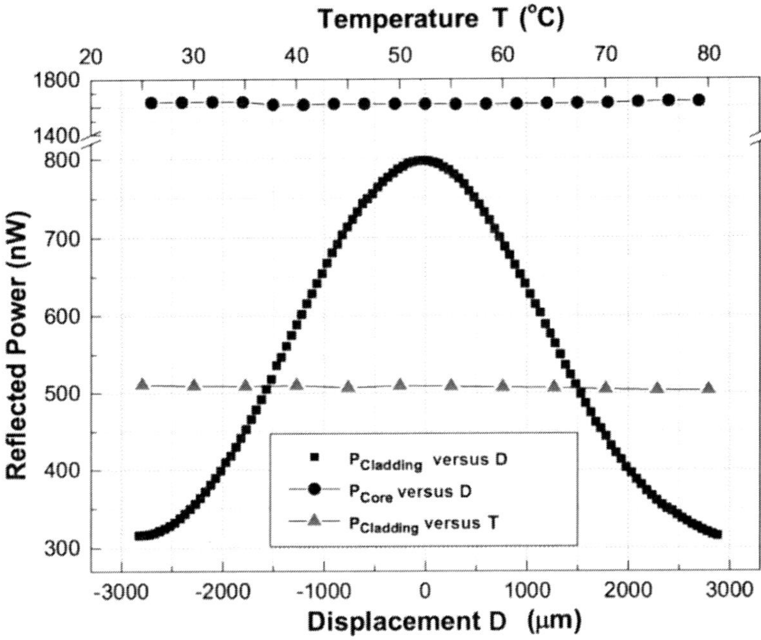

Figure 3. Cascade of up-taper and fiber Bragg grating (FBG) [54]. Behavior of cladding mode and core mode power versus displacement and cladding mode power versus temperature.

Gas or liquid flow measurement is required in the automotive industry, in material processing, in marine applications and in a number of other areas. Fiber optic technology has been largely employed [55,56,57,58,59]. For example, an optical fiber thermal anemometer based on a short silver-film-coated FBG and a core-offset fusion splice is proposed in [60]. The set-up is schematically shown in Figure 5. The core-offset splice is performed in order to couple most of the laser power from the fiber core to the cladding, see photograph and sketch of the core-offset splicing point. The silver coating close to the FBG absorbs the pump laser and generates heat. A pump laser at 1450 nm with a maximum output power of 450 mW has been used to heat the FBG. Therefore, the generated heat increases the temperature and thus the Bragg wavelength of the FBG. The anemometer is cooled down by the air flow through the anemometer. As a consequence, the Bragg wavelength shifts toward the short wavelength. The velocity of the air flow can be measured from the amount of wavelength shift of the FBG with respect to the nominal Bragg wavelength. In this way, a highly sensitive anemometer is obtained. A linear response of 45.3 pm/m/s and a high resolution of 0.022 m/s were achieved for an airflow velocity below 6 m/s. The sensor characteristic is shown in Figure 6, where the Bragg wavelength versus the airflow velocity is illustrated.

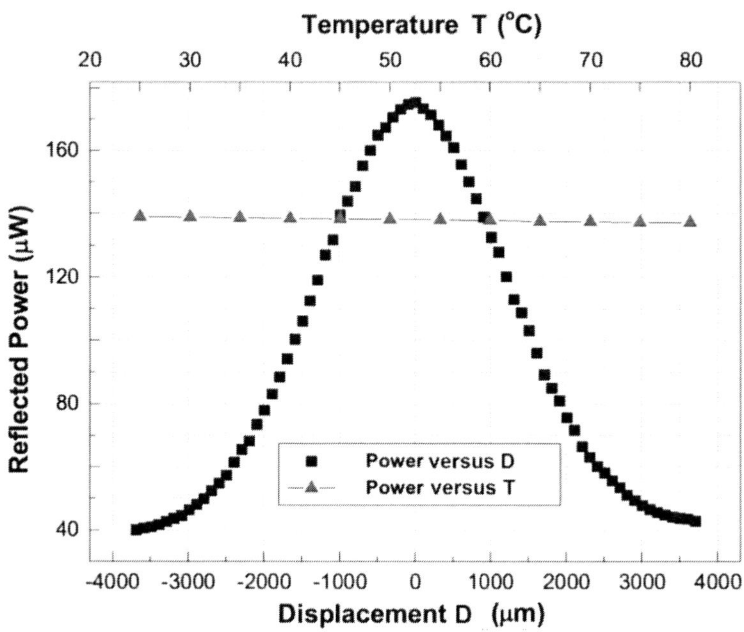

Figure 4. Cascade of up-taper and chirped FBG (CFBG) [54]. Behavior of cladding mode power and core mode power versus displacement and cladding mode power versus temperature.

Pressure sensors are desired in a number of areas e.g., industry, structural safety monitoring, oil wells, and power plants. Pressure sensors based on fiber Bragg gratings (FBGs) on side-hole fibers are very attractive [61,62,63,64]. In [65] a fiber-optic pressure sensor based on a π-phase-shifted fiber Bragg grating (πFBG) fabricated on a side-hole fiber is reported. A πFBG can be considered as aFabry-Perot (FP) resonator formed by two FBG mirrors. The side-hole fiber consists of two air channels running through the fiber cladding as shown in Figure 7. A hydrostatic pressure acting on the fiber surface induces an anisotropic stress distribution. Therefore a fiber birefringence via the photoelastic effect occurs. The fiber birefringence, proportional to the hydrostatic pressure, can be measured by evaluating the separation of the FBG peaks associated with the two polarization states. Therefore the FBG reflection spectrum gives a measurement of the pressure. The temperature can be simultaneously measured by monitoring the Bragg wavelength shift of the FBG peak. The performance can be enhanced by employing two πFBGs [65].

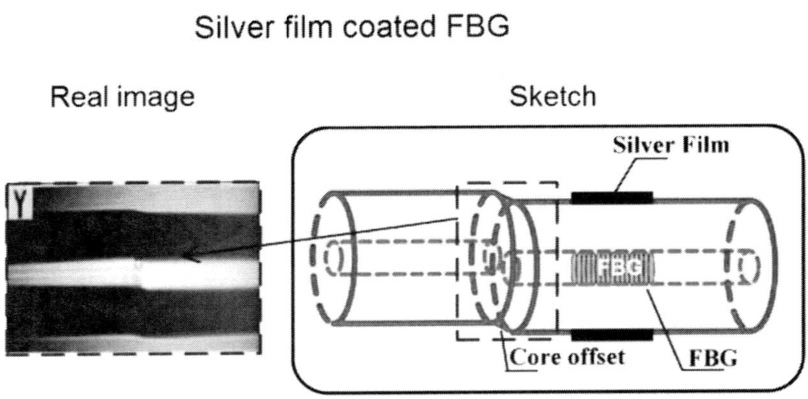

Figure 5. Schematic of the set-up proposed in [60]. FBG-based thermal anemometer.

The πFBG demonstrated in [65] is 8.3 mm long and the line width of each spectral notch is 3.6 pm, corresponding to a quality factor of 4.3×10^5. The spectral notch separation exhibited a sensitivity of 20 pm/kpsi to pressure. The Bragg wavelength shift exhibited a sensitivity of 11.4

pm/°C to temperature. A pressure detection limit of 1.4 psi and a temperature detection limit of 0.0025 °C were obtained.

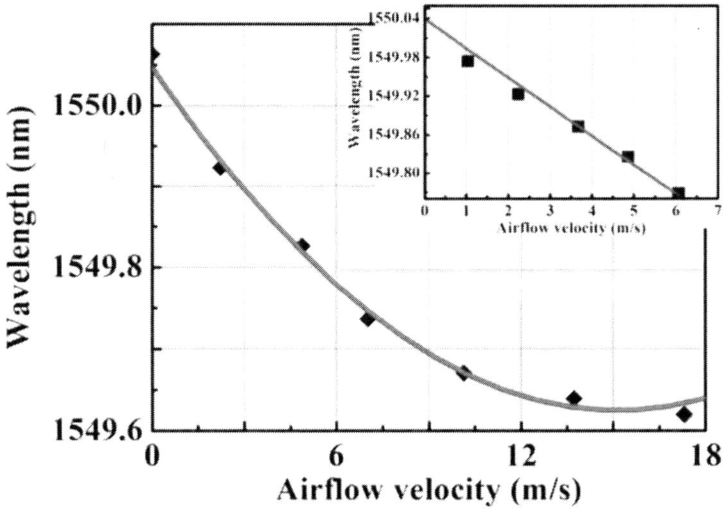

Figure 6. Resonant wavelength versus applied airflow velocity of the anemometer using silver-coated fiber Bragg grating [60].

2.2. Passive Long Period Grating Sensors

LPGs couple light very efficiently between two co-propagating modes in optical fibers. The resonance wavelength of an LPG, such as for the other gratings, is determined by the mode phase matching condition and by mode overlapping. It depends on the effective refractive index difference between these two coupled modes. Similarly to that occurring for FBGs, LPGs can be utilized to construct sensors for the detection of various physical parameters, including external refractive index, temperature, bending, loading, and liquid level [66,67]. LPG-based liquid-level sensor LLSs exhibit high sensitivity to external liquid (due to the light coupling to cladding modes) and high feasibility. In [68] an optical liquid-level sensor (LLS) based on long period fiber grating (LPG) interferometer was experimentally demonstrated. More precisely, an in-fiber Mach-Zehnder interferometer was obtained by employing two identical 3-dB LPGs. The sensing element, exposed to different liquids, is the fiber portion between two LPGs. The experimental sensor characteristic exhibited good linearity and a large measurement range. In Figure 8 the wavelength shift versus the liquid level is depicted for different refractive indices of the surrounding medium (liquids): 1.33 (rectangles), 1.398 (circles), and 1.413 (triangles) when the sensing element between two gratings is 250 mm long.

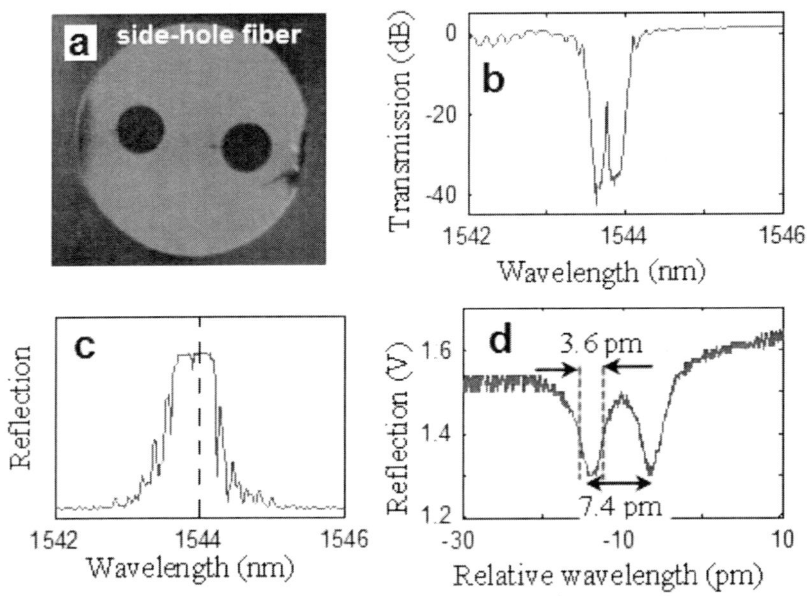

Figure 7. Fiber-optic pressure sensor based on a π-phase-shifted fiber Bragg grating (πFBG) fabricated on a side-hole fiber [65]. (**a**) Section of the side-hole fiber; (**b**) Transmission versus wavelength; (**c**) and (**d**) Reflection versus wavelength of a πFBG fabricated on the side-hole fiber.

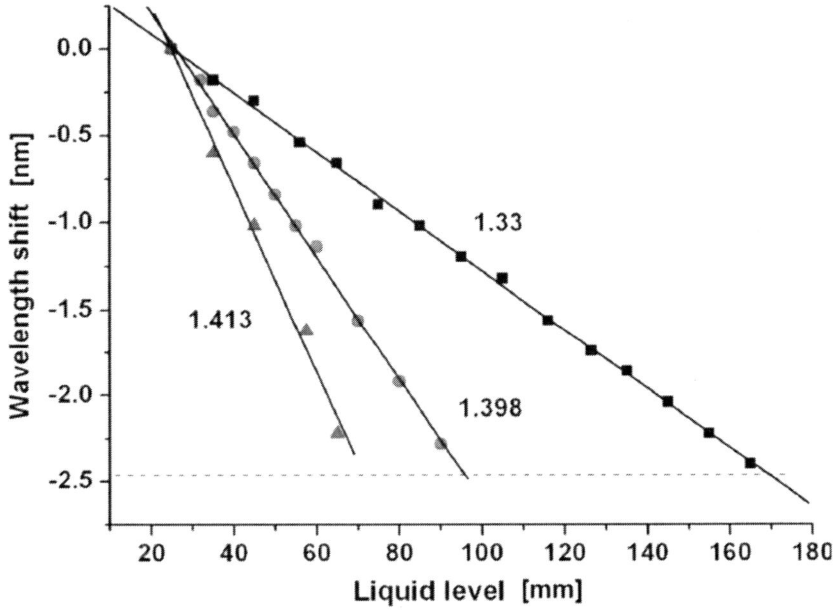

Figure 8. The relationship between peak wavelength and liquid-level with the surrounding liquid refractive indices [68].

LPGs such as FBGs can be utilized to enhance particular physical effects, e.g., Surface Plasmon Resonance (SPR). SPR is employed in a high number of sensors [69,70,71,72,73,74,75], it is based on electron density oscillations, propagating as transverse-magnetic (TM) polarized waves at the interface between a metal and a dielectric medium. These waves are highly attenuated evanescent ones. LPG can be employed to achieve SPR of a single cladding mode [76,77,78,79]. In [80] the fabrication of a fiber-optic refractive index (RI) sensor has been reported. A LPG has been fabricated to achieve surface plasmon resonance of a single cladding mode at the gold-coated tip of a single-mode fiber. The set-up is shown in Figure 9.

The sensor is based on a single-mode fiber. The fiber tip is covered by a gold-plate, this part is the sensing area which is close to core LPG. A thin poly-directional coating supports the formation of SPR cladding modes.

This structure has been proposed to obtain a biosensor. In fact, the power attenuation is strongly affected by the refractive index of the surrounding analyte medium. Small RI changes, e.g., caused by biomolecular binding processes, induce changes of the transmitted optical power. The LPG is designed for bidirectional coupling of the fundamental core mode with a cladding mode. Then the cladding modes are converted into an SPR cladding mode. The sensor has exhibited a high level of sensitivity in a miniaturized sensing area. LPGs with periods of 114 μm were fabricated to couple in to the designated cladding mode $HE_{1,20}$ at a wavelength of around 660 nm. LPG lengths of 30 mm enabled mode coupling of 15 dB.

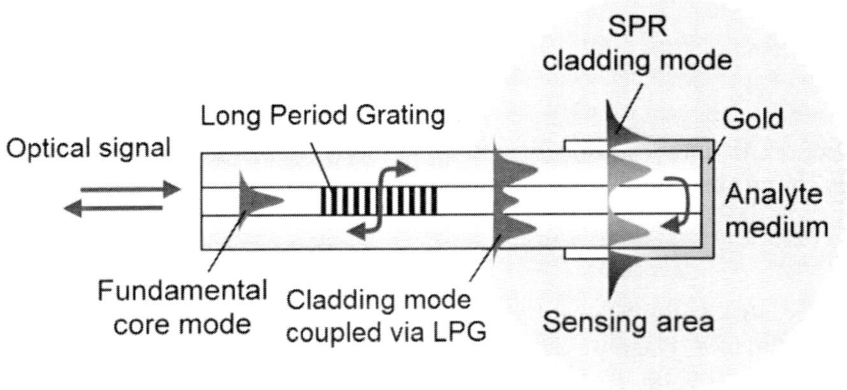

Figure 9. Set-up of the fiber-optic surface plasmon resonance (SPR) sensor proposed in [80].

Figure 10 shows the measured normalized transmission spectra for increasing refractive index of the surrounding medium (1–1.38). The LPG resonances are close to 660.6 nm. The vertical dashed line is the spectral location of the maximum SPR transmission loss.

LPGs have been inscribed in MOFs filled with ferrofluid. Figure 11 shows the microscopic photograph of the Ferrofluid-Infiltrated

Microstructured Optical Fiber Long-Period Grating reported in [81]. The cladding modes can be tuned more easily than the core mode. The mode dispersion curves are affected by the applied magnetic field. An enhancement of the applied magnetic field gradually reduces the refractive index of ferrofluid. This effect causes the decreasing of the effective refractive index difference between core and cladding modes, as shown in Figure 12. The MOF-based LPG has shown a sensitivity of 1.946 nm/Oe for a magnetic range of 0–300 Oe. The potential applications as a high-sensitivity magnetic-field sensor are apparent.

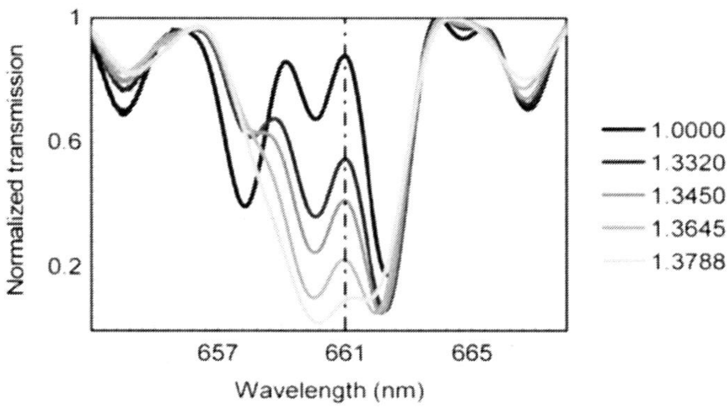

Figure 10. Normalized transmission spectra of the Miniaturized Long-Period Fiber Grating Assisted Surface Plasmon Resonance Sensor [80].

2.3. Active Long Period Grating Sensors

Laser sensors, include active behavior (e.g., due to the rare earth doped materials) and grating transmission spectra (e.g., FBG or LPG), which can be highly sensitive to the variation of some measurands. In [82,83] the design criterion to choose the suitable spatial distribution of multiple long period gratings MLPG couplers with respect to a MOF laser cavity was identified. The reported model extensively illustrates the behavior of the LPG in active medium, via the coupled mode equation and the rate equation of the rare earth ion population. In fact, it describes an Yb^{3+}-doped MOF laser cavity by taking into account the presence of different MLPGs inscribed within the core. The MLPGs reported in [83] enable the interaction among the two-fold degenerate fundamental mode guided in the fiber core (HE_{11}) and the inner cladding modes, at the pump wavelength. The model includes the following kinds of couplers: (a) MLPGs-P, Multiple Long Period Gratings in Passive fibers, i.e., grating cascades inscribed in the undoped region, just outside the laser cavity; (b) MLPGs-A, Multiple Long Period Gratings in Active fibers, i.e., grating cascades inscribed within the laser cavity (see Figure 13); (c) DS-MLPGs, Double Stage Multiple Long Period Gratings, i.e., a combination of passive grating cascades, inscribed outside the laser cavity, and active ones, inscribed within the laser cavity, are employed). Figure 14 shows the power in the cladding at pump wavelength $P_{clad}(\lambda_p)$, in the core at pump wavelength $P_{core}(\lambda_p)$, forward and backward in the core at signal wavelength $P^+_{core}(\lambda_s)$, $P^-_{core}(\lambda_s)$ versus the position along the fiber longitudinal axis z for the DS-MLPGs. The vertical lines represent the grating cascade along the fiber axis.

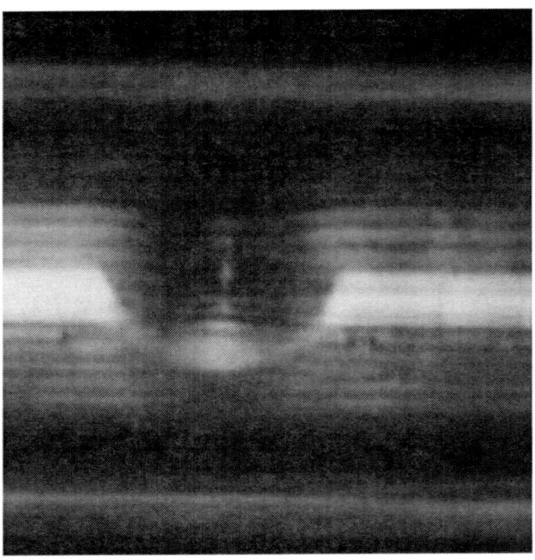

Figure 11. LPG inscribed in microstructured optical fiber (MOF) filled with ferrofluid [81].

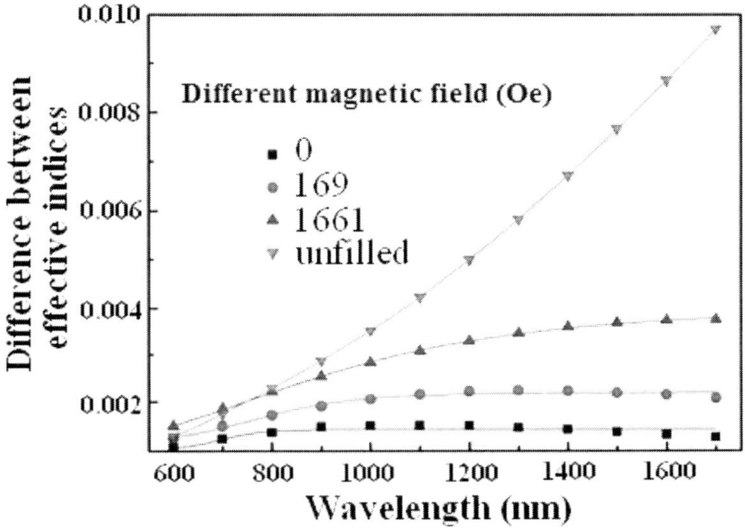

Figure 12. Difference between cladding-mode and core-mode effective indices for different applied magnetic fields [81].

Even if MLPGs-P, MLPGs-A and DS-MLPGs were designed in [83] to increase the pump power transfer from the inner cladding modes at the pump power towards the fundamental core mode of the MOF laser, the model can be employed for designing active MLPG sensors since both (i) the LPG transmission spectrum; and (ii) its efficiency as power coupler can be affected by different measurands (e.g., bending or strain).

Since the phase matching condition of the coupled modes in the LPG is strongly affected by small changes in the effective index difference, LPG refractive index sensors are feasible. It is well known that by filling the holes of MOFs with some functional material, their modal and dispersion properties may change strongly. This also affects the tuning properties of LPGs inscribed in such fibers [81]. Therefore, MOF based LPGs are highly sensitive/tunable devices [84,85,86].

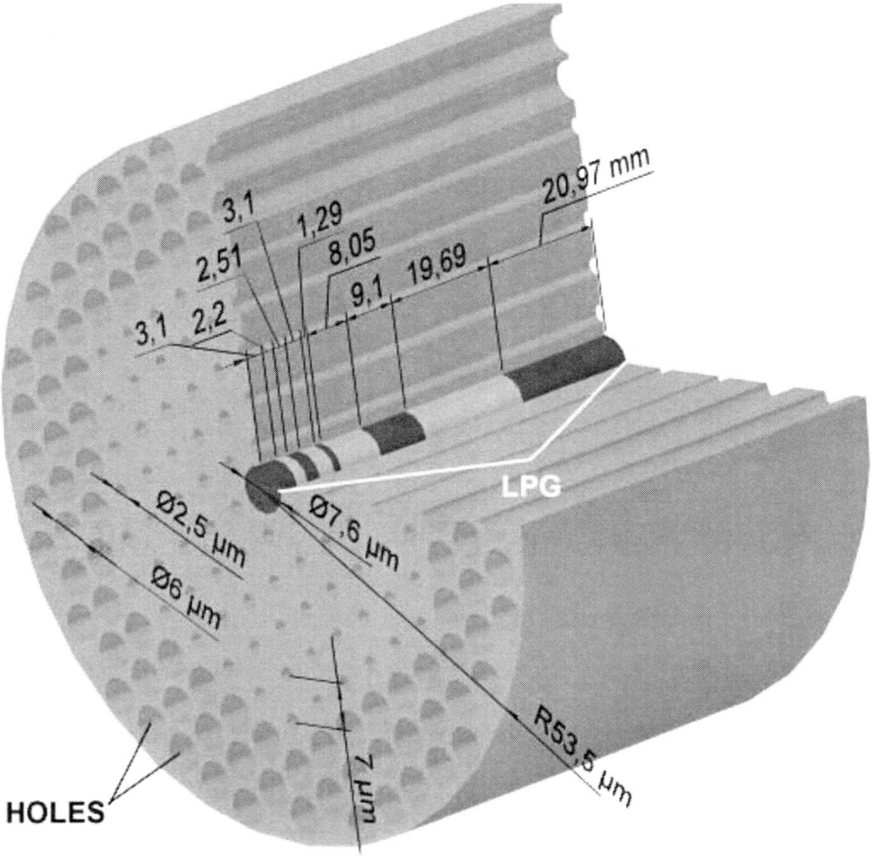

Figure 13. 3D sketch of the ytterbium-doped MOF laser with the MLPGs-A grating cascade [83].

Recently LPGs, similarly to FBG, were proposed for optical pulse shaping and coding techniques [87,88,89,90,91]. These topics are interesting within communication systems and, in addition, to develop a large interrogating optical network e.g., based on Optical Time Domain Reflectometer (OTDR) techniques and communication protocols.

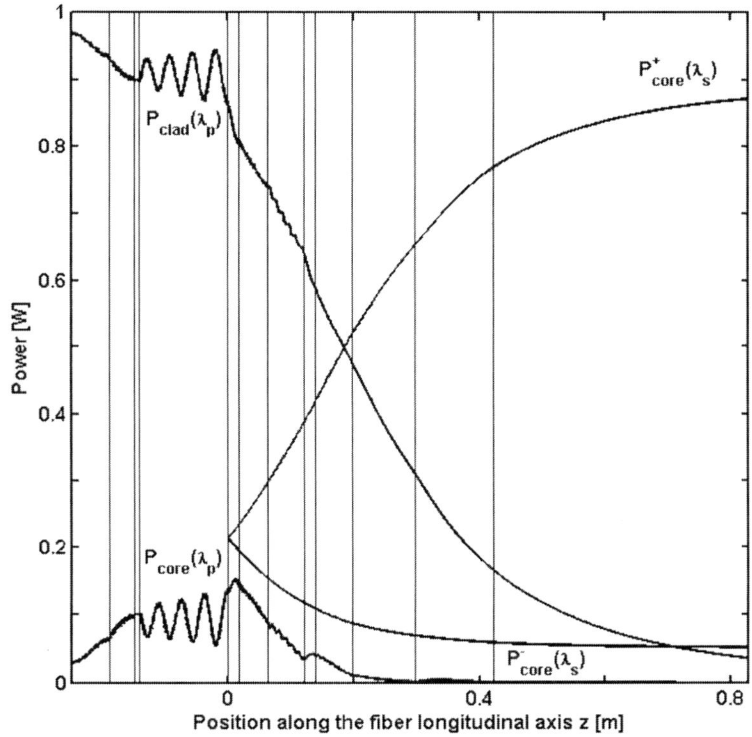

Figure 14. Power $P_{clad}(\lambda_p)$, $P_{core}(\lambda_p)$, $P^{+}_{core}(\lambda_s)$, $P^{-}_{core}(\lambda_s)$ versus the position along the fiber longitudinal axis z, DS-MLPGs [83].

3. EVANESCENT FIELD, HOLLOW CORE, LIQUID FILLED AND EXPOSED OPTICAL FIBER SENSORS

Optical fiber evanescent wave sensors are a valid alternative for chemical analysis [92]. Distributed optical fiber sensors based on evanescent field interactions enable the measurement of chemical distributions over very long distances. Their potential applications are in structural health monitoring, environmental monitoring, homeland security, and many other applications [93,94,95,96,97,98,99,100].

The optimization of evanescent field optical sensors plays an important role in the development of low-cost pollution monitoring. In this case, the detection is obtained via transmittance or absorption measurements at the wavelength where the contaminant exhibits the highest absorption peak. In [101,102,103,104,105,106,107] optical sensing of hydrocarbons in air or in water, using UV or NIR evanescent field optical sensors was reported. Figure 15 illustrates the example of the set-up reported in [106].

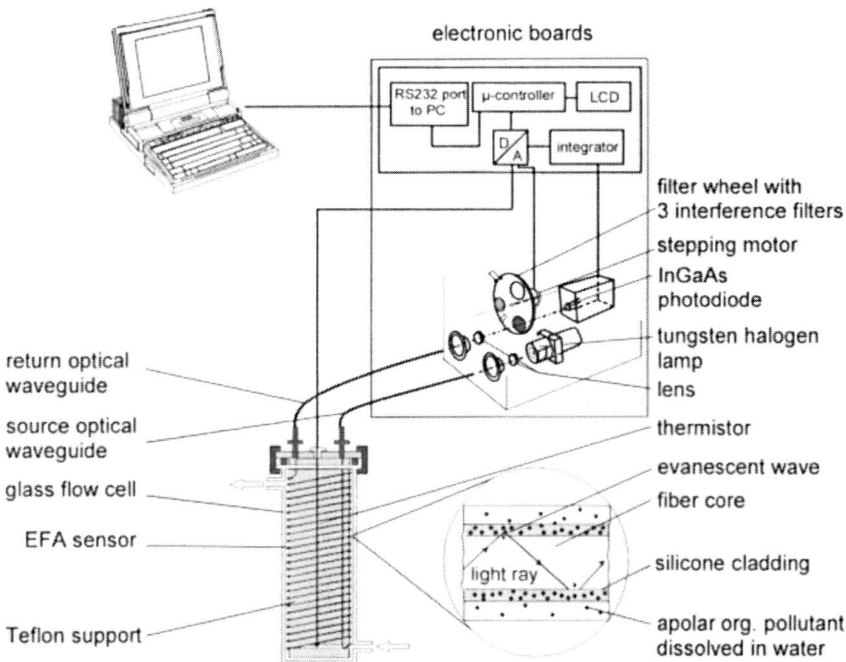

Figure 15. Evanescent field absorption sensing principle and instrumental set-up of coiled fiber-optic sensor installed in a flow cell and and NIR bandpass filter photometer unit [106].

In [108,109,110] fiber optic sensors for biosensing are reported. In particular in [108] the design of a photonic crystal fiber sensor for methadone detection in water was performed. The sensor feasibility was numerically investigated by employing a Finite Element Method (FEM) numerical code. The fiber cross section, including a polymeric cladding as sensitive layer, was optimized. In the design, the thickness of the polymeric layer, the hole diameter, the hole-to-hole spacing and the transversal distribution of the holes were parametrically varied. The design of the fiber geometry was performed to balance competing requirements such as the reduction of the number of the propagating modes and the improvement of the evanescent field optical power interacting with the sensitive layer. The polymeric sensitive layer interacts with the water polluted by the methadone. At equilibrium, the contaminant concentration (C_p) in the polymer and in the water (C_w) are linked by the distribution constant $K_{p/w} = C_p/C_w$, whose value is peculiar to that of the used polymer. The methadone induces the change of the polymer absorption loss which increases by increasing the contaminant concentration. Therefore, a change of the imaginary part of the complex effective refractive index of the guided modes occurs. In particular, the light guided in the fiber core suffers a power attenuation due to the methadone. The optimal UV operation wavelength is that at which the methadone absorption peak occurs. Figure 16 shows the simulated transverse profile of the electric field of the HE_{11} guided propagation mode (a) for square-lattice; and (b) for the hexagonal-lattice photonic crystal fibers.

Figure 17 depicts the absorbance A versus the methadone concentration for the square-lattice photonic crystal fibers (PCF) (full curve), hexagonal-lattice PCF (dash curve) and conventional step-index fiber (dot curve) having core radius R = 8 μm and PDMS thickness t_c = 0.4 μm, for the sensor length L = 15 cm.

In [109] the design of a silica exposed-core fiber sensor for methadone detection in water was reported; exposed-core fiber coated with polydimethylsiloxane PDMS was fabricated at the University of Adelaide in order to further enhance the sensitivity. Figure 18 reports the sketch of (a) exposed-core fiber; and (b) exposed-core fiber with PDMS coating. By simulation the absorbance sensitivity for a fiber length L = 10 cm for the exposed core fiber without PDMS is close to $A = 0.3 \times 10^{-6}$ [ppb−1] while it is close to A = 0.011 [ppb−1] for exposed core fiber with PDMS. Lead silicate glass (F2) was considered in the simulation as an alternative to silica glass and different contaminants (benzene, toluene etc.).

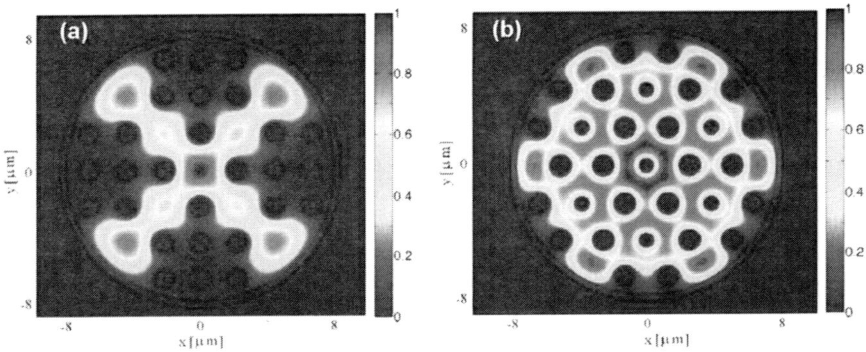

Figure 16. Simulated transverse profile of the electric field of the HE_{11} guided propagation mode (**a**) for square-lattice; and (**b**) for the hexagonal-lattice photonic crystal fibers [108].

The air holes of ad-hoc designed MOFs can be filled with different materials [111,112,113]. Hollow-core MOFs having their core filled with gas or liquids were investigated for sensing and spectroscopic applications [114,115,116]. In [117] a liquid-core air-clad microstructured fiber was developed for determination of ethanol and sucrose concentrations in aqueous solutions by Raman spectroscopy. The application of the sensor to white wine characterization has been proposed. Figure 19 shows the SEM images of the fiber (left) and of the narrow silica bridges forming the air-clad (right) employed in [117].

Figure 20 shows the measured and reproduced Raman spectra of wine with the spectral contribution of water, ethanol, and sucrose.

By considering the optical fiber as a part of the sensing mechanism a number of other devices have been proposed. As further examples, passive and active microspheres coupled to taper fibers were proposed for biosensing applications [118,119,120,121,122,123,124, 125,126, 127,128,129,130]. To conclude, the variety of e proposed optical fiber sensors as well as their potential is extremely impressive.

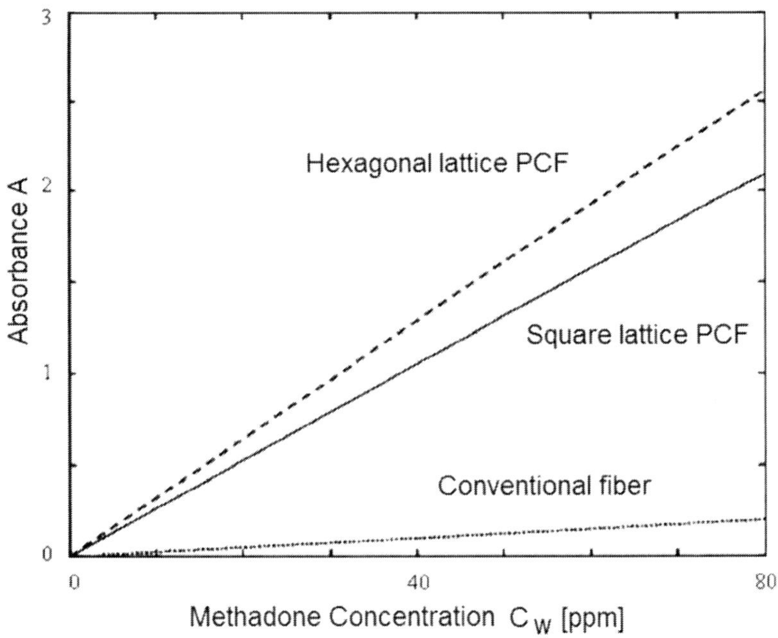

Figure 17. Absorbance A versus the methadone concentration for the square-lattice photonic crystal fibers (PCF) (full curve), hexagonal-lattice PCF (dash curve) and conventional step-index fiber (dot curve) [108].

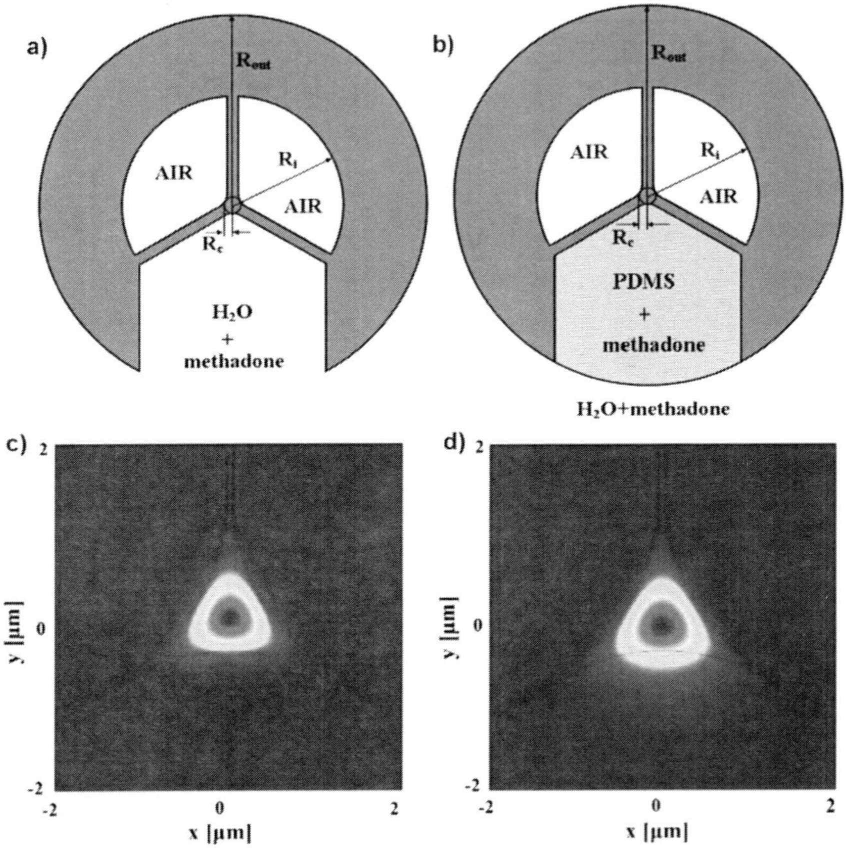

Figure 18. Sketch of the (**a**) exposed-core fiber; and (**b**) exposed-core fiber with polydimethylsiloxane PDMS coating. Simulated transverse electric field norm of the HE_{11} y-polarized guided mode at the wavelength $\lambda = 292$ nm for (**c**) exposed-core fiber; and (**d**) exposed-core fiber with PDMS coating [109].

4. CONCLUSIONS

FBG, LPG, evanescent field and hollow core optical fiber sensors are briefly illustrated here. More precisely, examples of recent optical fiber sensors for the measurement of strain, temperature, displacement, air flow, pressure, liquid-level, magnetic field, as well as the determination of methadone, hydrocarbons, ethanol, and sucrose are discussed.

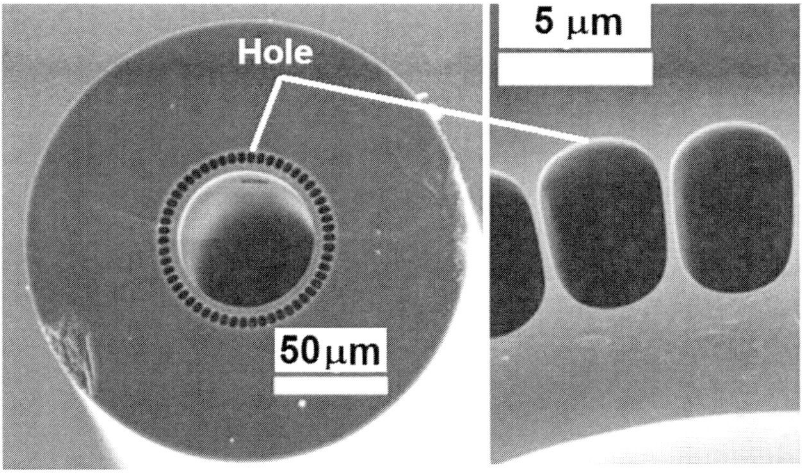

Figure 19. SEM images of the fiber (left) and of the narrow silica bridges forming the air-clad (right) [117].

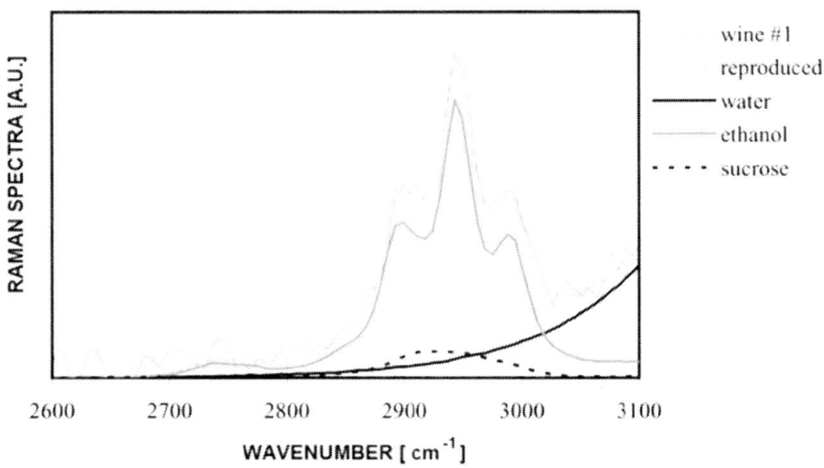

Figure 20. Raman spectra of wine [117].

CONFLICTS OF INTEREST

The authors declare no conflict of interest.

REFERENCES

1. Girard, S.; Mescia, L.; Vivona, M.; Laurent, A.; Ouerdane, Y.; Marcandella, C.; Prudenzano, F.; Boukenter, A; Robin, T.; Paillet, P; Goiffon, V.; Gaillardin, M.; Cadier, B.; Pinsard, E.; Cannas, M.; Boscaino, R. Design of radiation-hardened rare-earth doped amplifiers through a coupled experiment/simulation approach. J. Lightwave Technol. **2013**, 31, 1247–1254.

2. Lee, B. Review of the present status of optical fiber sensors. Opt. Fiber Technol. **2003**, 9, 57–79.

3. Pelli, S.; Chiasera, A.; Ferrari, M.; Righini, G.C. Spectroscopic Techniques for Sensors. In An Introduction to Optoelectronic Sensors; Series in Optics and Photonics; World Scientific Publishing Co. Pte. Ltd.: Singapore, 2009; Volume 7, ISBN: 978-981-283-412-6 981-283-412-5.

4. Cordeiro, C.M.B.; Franco, M.A.R.; Chesini, G.; Barretto, E.C.S.; Lwin, R.; Cruz, C.H.B.; Large, M.C.J. Microstructured-core optical fibre for evanescent sensing applications. Opt. Express **2006**, 14, 13056–13066.

5. Jensen, J.; Hoiby, P.; Emiliyanov, G.; Bang, O.; Pedersen, L.H.; Bjarklev, A. Selective detection of antibodies in microstructured polymer optical fibers. Opt. Express **2005**, 13, 5883–5889.

6. Jensen, J.B.; Pedersen, L.H.; Hoiby, P.E.; Nielsen, L.B.; Hansen, T.P.; Folkenberg, J.R.; Riishede, J.; Noordegraaf, D.; Nielsen, K.; Carlsen, A.; Bjarklev, A. Photonic crystal fiber based evanescent-wave sensor for detection of biomolecules in aqueous solutions. Opt. Lett. **2004**, 29, 1974–1976.

7. Lehmann, H.; Brückner, S.; Kobelke, J.; Schwotzer, G.; Schuster, K.; Willsch, R. Toward photonic crystal fiber based distributed chemosensors. Proc. SPIE **2005**, 5855, 419–422.

8. Rothwell, J.H.; Flavin, D.A.; MacPherson, W.N.; Jones, J.D.C.; Knight, J.C.; Russell, P.St.J. Photonic sensing based on variation of propagation properties of photonic crystal fibres. Opt. Express **2006**, 14, 12445–12450.

9. Ma, J; Bock, W.J.; Wang, Z. Y.; Hao, W; MacKinnon, S. M. Fiber-optic membrane fluorescent sensor based on photonic crystal fiber with a glass rod in the fiber end. In IEEE Sensors 2005, Irvine, CA, USA, 30 October–November 2005; pp. 1096–1099.

10. Cox, F.M.; Argyros, A.; Large, M.C.J. Liquid-filled hollow core microstructured polymer optical fiber. Opt. Express **2006**, 14, 4135–4140.

11. Fini, J.M. Microstructure fibres for optical sensing in gases and liquids. Meas. Sci. Technol. **2004**, 15, 1120–1128.

12. Monro, T.M.; Belardi, W.; Furusawa, K.; Baggett, J.C.; Broderick, N.G.R.; Richardson, D.J. Sensing with microstructured optical fibres. Meas. Sci. Technol. **2001**, 12, 854–858.

13. Hill, K.O.; Meltz, G. Fiber Bragg grating technology fundamentals and overview. J. Lightwave Technol. **1997**, 15, 1263–1276.

14. Hill, K.O.; Fujii, Y.; Johnson, D.C.; Kawasaki, B.S. Photosensitivity in optical fiber waveguides: Application to reflection filter fabrication. Appl. Phys. Lett. **1978**, 32, 647–649.

15. Culshaw, B.; Kersey, A. Fiber-optic sensing: A historical perspective. J. Lightwave Technol. **2008**, 26, 1064–1078.

16. Guan, B.; Jin, L.; Zhang, Y.; Tam, H. Polarimetric heterodyning fiber grating laser sensors. J. Lightwave Technol. **2012**, 30, 1097–1111.

17. Dong, L.; Loh, W.H.; Caplen, J.E.; Minelly, J.D.; Hsu, K.; Reekie, L. Efficient single-frequency fiber lasers with novel photosensitive Er/Yb optical fibers. Opt. Lett. **1997**, 22, 694–696.

18. Hill, D.J.; Hodder, B.; DeFreitas, J.; Thomas, S.D.; Hickey, L. DFB Fibre-Laser Sensor Developments. In Proceedings of SPIE, the 17th International Conference on Optical Fibre Sensors, Bruges, Belgium, 23 May 2005; Volume 5855, pp. 904–907.

19. Li, J.; He, F.; Xu, T.; Wang, Y.; Liu, Y. High Performance Distributed Feedback Fiber Laser Sensor Array System. In Proceedings of Asia Communications and Photonics Conference Exhibition, Shanghai, China, 2–6 November 2009; Volume 7634, pp. 76340K:1–76340K:9.

20. Cranch, G.A.; Flockhart, G.; Kirkendall, C.K. Distributed feedback fiber laser strain sensors. IEEE Sens. J. **2008**, 8, 1161–1172.

21. Beverini, N.; Maccioni, E.; Morganti, M.; Stefani, V.; Falciai, R.; Trono, C. Fiber laser strain sensor device. J. Opt. A: Pure Appl. Opt. **2007**, 9, 958–962.

22. Saitoh, K.; Koshiba, M.; Hasegawa, T.; Sasaoka, F. Chromatic dispersion control in photonic crystal fibers: Application to ultra-flattened dispersion. Opt. Express **2003**, 11, 843–852.

23. Westbrook, P.S.; Eggleton, B.J.; Windeler, R.S.; Hale, A.; Strasser, T.A.; Burdge, G.L. Cladding-mode resonances in hybrid polymer-silica microstructured optical fiber gratings. IEEE Photonics Technol. Lett. **2000**, 12, 495–497.

24. Lee, B.H.; Liu, Y.; Lee, S.B.; Choi, S.S.; Jang, J.N. Displacements of the resonant peaks of a long-period fiber grating induced by a change of ambient refractive index. Opt. Lett. **1997**, 22, 1769–1771.

25. Steivurzel, P.; Moore, E.D.; Mägi, E.C.; Kuhlmey, B.T.; Eggleton, B.J. Long period grating resonances in photonic bandgap fiber. Opt. Express **2006**, 14, 3007–3014.

26. Lee, D.; Jung, Y.; Jeong, Y.S.; Oh, K.; Kobelke, J.; Schuster, K.; Kirchhof, J. Highly polarization-depend periodic coupling in mechanically induced long period grating over air-silica fibers. Opt. Lett. **2006**, 31, 296–298.

27. Lim, J.H.; Lee, K.S.; Kim, J.C.; Lee, B.H. Tunable fiber gratings fabricated in photonic crystal fiber by use of mechanical pressure. Opt. Lett. **2004**, 29, 331–333.

28. Petrovic, J.S.; Dobb, H.; Mezentsev, V.K.; Kalli, K.; Webb, D.J.; Bennion, I. Sensitivity of LPGs in PCFs fabricated by an electric Arc to temperature, strain, and external refractive index. J. Lightwave Technol. **2007**, 25, 1306–1312.

29. Zhao, C.L.; Xiao, L.; Ju, J.; Demokan, M.S.; Jin, W. Strain and temperature characteristics of a long-period grating written in a photonic crystal fiber and its application as a temperature-insensitive strain sensor. J. Lightwave Technol. **2008**, 26, 220–227.

30. Eggleton, B.J.; Westbrook, P.S.; Windeler, R.S.; Spalter, S.; Strasser, T.A. Grating resonances in air-silica microstructured optical fibers. Opt. Lett. **1999**, 24, 1460–1462.

31. Erdogan, T. Cladding-mode resonances in short-and long-period fiber grating filters. J. Opt. Soc. Am. A **1997**, 14, 1760–1773.

32. Mescia, L. Design of long-period gratings in cladding-pumped microstructured optical fiber. J. Opt. Soc. Am. B **2008**, 25, 1883–1839.

33. Calò, G.; D'Orazio, A.; De Sario, M.; Mescia, L.; Petruzzelli, V.; Allegretti, L.; Palmisano, T.; Prudenzano, F. Improvement of the Pump Power Coupling in Double Cladding Photonic Crystal Fiber. In Proceedings of IEEE/LEOS Winter Topical Meeting Series, Sorrento, Italy, 14–16 January 2008; pp. 146–147.

34. Baek, S.; Roh, S.; Jeong, Y.; Lee, B. Experimental demonstration of enhancing pump absorption rate in cladding-pumped ytterbium-

doped fiber laser using pump-coupling long-period gratings. IEEE Photonics Technol. Lett. **2006**, 18, 700–702.

35. Allsop, T.; Kalli, K.; Zhou, K.; Smith, G.; Laia, V.; Smith, G.; Dubov, M.; Webb, D.; Bennion, V. Long period gratings written into a photonic crystal fibre by a femtosecond laser as directional bend sensors. Opt. Commun. **2008**, 281, 5092–5096.

36. Long, J.; Zhi, W.; Yange, L.; Guiyun, K.; Xiaoyi, D. Ultraviolet-inscribed long period gratings in all-solid photonic bandgap fibers. Opt. Express **2008**, 16, 21119–21131.

37. Shujing, L.; Long, J.; Wei, J.; Dongning, W.; Changrui, L.; Ying, W. Structural long period gratings made by drilling micro-holes in photonic crystal fibers with a femtosecond infrared laser. Opt. Express **2010**, 18, 5496–5503.

38. D'Orazio, A.; De Sario, M.; Mescia, L.; Petruzzelli, V.; Prudenzano, F. Design of double-clad ytterbium doped microstructured fibre laser. Appl. Surf. Sci. **2005**, 248, 499–502.

39. Prudenzano, F. Erbium-doped hole-assisted optical fiber amplifier: Design and optimization. J. Lightwave Technol. **2005**, 23, 330–340.

40. Sabaiean, M.; Nadgaran, H.; De Sario, M.; Mescia, L.; Prudenzano, F. Thermal effects on double clad octagonal Yb: glass fiber laser. Opt. Mater. **2009**, 31, 1300–1305.

41. Prudenzano, F.; Mescia, V.; Allegretti, L.; De Sario, M.; Palmisano, T.; Smektala, F.; Moizan, V.; Nazabal, V.; Troles, J. Design of Er^{3+}-doped chalcogenide glass laser for MID-IR application. J. Non-Cryst. Solids **2009**, 355, 1145–1148.

42. Li, L.; Schülzgen, A.; Temyanko, V.L.; Morrell, M.M.; Sabet, S.; Li, H.; Moloney, J.V.; Peyghambarian, N. Ultracompact cladding-pumped 35-mm-short fiber laser with 4.7-W single-mode. Applied Physics Letters. **2006**, 88, 161106:1–161106:2.

43. Prudenzano, F.; Mescia, L.; Allegretti, L.; Moizan, V.; Nazabal, V.; Smektala, F. Theoretical study of cascade laser in erbium-doped chalcogenide glass fibers. Opt. Mater. **2010**, 33, 241–245.

44. Mescia, L.; Smektala, F.; Prudenzano, F. New trends in amplifiers and sources via chalcogenide photonic crystal fibers. Int. J. Opt. **2012**, 2012, E212091.

45. Prudenzano, F.; Mescia, L.; Di Tommaso, A.; Surico, M.; De Sario, M. Design and refinement of rare earth doped multicore fiber lasers. Opt. Mater. **2013**, 35, 1941–1946.

46. Rajan, G.; Ramakrishnan, M.; Semenova, Y.; Milenko, K.; Lesiak, P.; Domanski, A.W.; Wolinski, T.R.; Farrell, G. A photonic crystal fiber and fiber Bragg grating-based hybrid fiber-optic sensor. Syst. IEEE Sens. J. **2012**, 12, 39–43.

47. Zou, Y.; Dong, X.; Lin, G.; Adhami, R. Wide range FBG displacement sensor based on twin-core fiber filter. J. Lightwave Technol. **2012**, 30, 337–343.

48. Zhu, Y.; Shum, P.; Lu, C.; Lacquet, M.; Swart, P.; Chtcherbakov, A.; Spammer, S. Temperature insensitive measurements of static displacements using a fiber Bragg grating. Opt. Express **2003**, 11, 1918–1924.

49. Guo, T.; Shao, L.; Tam, H.-Y.; Krug, A.; Albert, J. Tilted fiber grating accelerometer incorporating an abrupt biconical taper for cladding to core recoupling. Opt. Express **2009**, 17, 20651–20660.

50. Shao, L.Y.; Albert, J. Compact fiber-optic vector inclinometer. Opt. Lett. **2010**, 35, 1034–1036.

51. Zhou, W.; Zhou, Y.; Dong, X.; Shao, L.-Y.; Cheng, J.; Albert, J. Fiber-optic curvature sensor based on cladding-mode Bragg grating excited by fiber multimode interferometer. IEEE Photonics J. **2012**, 4, 1051–1057.

52. Rong, Q.; Qiao, X.; Zhang, J.; Wang, R.; Hu, M.; Feng, Z. Simultaneous measurement for displacement and temperature using fiber Bragg grating cladding mode based on core diameter mismatch. J. Lightwave Technol. **2012**, 30, 1645–1650.

53. Gouveia, C.; Jorge, P.; Baptista, J.M.; Frazao, O. Temperature-independent curvature sensor using FBG cladding modes based on a core misaligned splice. IEEE Photonics Technol. Lett. **2011**, 23, 804–806.

54. Qi, T.; Xiao, S.; Shi, J.; Yi, L.; Zhou, Z.; Bi, M.; Hu, W. Cladding-mode backward-recoupling-based displacement sensor incorporating fiber up taper and Bragg grating. IEEE Photonics J. **2013**, 5, 7100608.

55. Zhao, Y.; Chen, K.; Yang, J. Novel target type flowmeter based on a differential fiber Bragg grating sensor. Measurement **2005**, 38, 230–235.

56. Jewart, C.; McMillen, B.; Cho, S.K.; Chen, K.P. X-probe flow using self-powered active fiber Bragg gratings. Sens. Actuators A Phys. **2006**, 127, 63–68.

57. Byrne, G.D.; James, S.W.; Tatam, R.P. A Bragg grating based fibre optic reference beam laser Doppler anemometer. Meas. Sci. Technol. **2001**, 12, 909–913.

58. Lien, V.; Vollmer, F. Microfluidic flow rate detection based on integrated optical fiber cantilever. Lab Chip **2007**, 7, 1352–1356.

59. Lee, C.L.; Hong, W.Y.; Hsieh, H.J.; Weng, Z.Y. Air gap fiber Fabry-Pe´ rot interferometer for highly sensitive microairflow sensing. IEEE Photonics Technol. Lett. **2011**, 23, 905–907.

60. Dong, X.; Zhou, Y.; Zhou, W.; Cheng, J.; Su, Z. Compact anemometer using silver-coated fiber Bragg grating. IEEE Photonics J. **2012**, 5, 1381–1386.

61. Kreger, S.; Calvert, S.; Udd, E. High Pressure Sensing Using Fiber Bragg Gratings Written in Birefringent Side Hole Fiber. In Proceedings of the 15th Optical Fiber Sensors Conference Technical Digest, Portland, OR, USA, 10 May 2002; pp. 355–358.

62. Chen, T.; Chen, R.; Jewart, C.; Zhang, B.; Cook, K.; Canning, J.; Chen, K.P. Regenerated gratings in air-hole microstructured fibers for high-temperature pressure sensing. Opt. Lett. **2011**, 36, 3542–3544.

63. Jewart, C.M.; Wang, Q.; Canning, J.; Grobnic, D.; Mihailov, S.J.; Chen, K.P. Ultrafast femtosecond-laser-induced fiber Bragg gratings in air-hole microstructured fibers for high-temperature pressure sensing. Opt. Lett. **2010**, 35, 1443–1445.

64. Chmielewska, E.; Urbanczyk, W.; Bock, W.J. Measurement of pressure and temperature sensitivities of a Bragg grating imprinted in a highly birefringent side-hole fiber. Appl. Opt. **2003**, 42, 6284–6291.

65. Zhang, Q.; Liu, N.; Fink, T.; Li, H.; Peng, W.; Han, M. Fiber-optic pressure sensor based on π-phase-shifted fiber Bragg grating on side-hole fiber. IEEE Photonics Technol. Lett. **2012**, 24, 1519–1522.

66. Khaliq, S.; James, S.W.; Tatam, R.P. Fiber-optic liquid-level sensor using a long-period grating. Opt. Lett. **2001**, 26, 1224–1226.

67. Yun, B.; Chen, N.; Cui, Y. Highly sensitive liquid-level sensor based on etched fiber Bragg grating. IEEE Photonics Technol. Lett. **2007**, 19, 1747–1749.

68. Fu, H.; Shu, X.; Zhang, A.; Liu, W.; Zhang, L.; He, S.; Bennion, I. Implementation and characterization of liquid-level sensor based on a long-period fiber grating mach-zehnder interferometer. IEEE Sens. J. **2011**, 11, 2878–2883.

69. Homola, J. Present and future of surface plasmon resonance biosensors. Anal. Bioanal. Chem. **2003**, 377, 528–539.

70. Räther, H. Surface Plasmons on Smooth and Rough Surfaces and on Gratings. In Springer Tracts in Modern Physics; Springer-Verlag: Berlin, Germany, 1988; Volume 11.

71. Kretschmann, E. Die Bestimmung optischer Konstanten vonMetallen durch anregung von oberflächenplasmaschwingungen. Zeitschrift für Physik A: Hadrons and Nuclei **1971**, 241, 313–324.

72. Dostáleka, J.; Čtyrokýa, J.; Homolaa, J.; Bryndab, E.; Skalskýa, M.; Nekvindovác, P.; Špirkovác, J.; Škvord, J.; Schröfele, J. Surface plasmon resonance biosensor based on integrated optical waveguide. Sens. Actuators B Chem. **2001**, 76, 8–12.

73. Gupta, B.D.; Verma, R.K. Surface plasmon resonance-based fiber optic sensors: Principle, probe designs, and some applications. J. Sens. **2009**.

74. Obando, L.; Gentleman, D.; Holloway, J.; Booksh, K. Manufacture of robust surface plasmon resonance fiber optic based dip-probes. Sens. Actuators B Chem. 2004, 100, 439–449.

75. Esteban, Ó.; Díaz-Herrera, N.; Navarrete, M.; González-Cano, A. Surface plasmon resonance sensors based on uniform-waist tapered fibers in a reflective configuration. Appl. Opt. 2006, 45, 7294–7298.

76. He, Y.J.; Lo, Y.L.; Huang, J.F. Optical-fiber surface plasmon-resonance sensor employing long period fiber gratings in multiplexing. J. Opt. Soc. Am. B 2006, 23, 801–811.

77. Spacáková, B.; Homola, J. Theoretical analysis of a fiber optic surface plasmon resonance sensor utilizing a Bragg grating. Opt. Express 2009, 17, 23254–23264.

78. Lu, Y.C.; Huang, W.P.; Jian, S.S. Influence of mode loss on the feasibility of grating-assisted optical fiber surface plasmon resonance refractive index sensors. J. Lightwave Technol. 2009, 27, 4804–4808.

79. Caucheteur, C.; Shevchenko, Y.; Shao, L.-Y.; Wuilpart, M.; Albert, J. High resolution interrogation of tilted fiber grating SPR sensors from polarization properties measurement. Opt. Express 2011, 19, 1656–1664.

80. Schuster, T.; Herschel, R.; Neumann, N.; Schäffer, C.G. Miniaturized long-period fiber grating assisted surface Plasmon resonance sensor. J. Lightwave Technol. 2012, 30, 1003–1008.

81. Miao, Y.; Zhang, K.; Liu, B.; Lin, W.; Zhang, H.; Lu, Y.; Yao, J. Ferrofluid-infiltrated microstructured optical fiber long-period grating. IEEE Photonics Technol. Lett. 2013, 25, 306–309.

82. Mescia, L.; Palmisano, T.; Surico, M.; Prudenzano, F. Long-period gratings for the optimization of cladding-pumped microstructured optical fiber laser. Opt. Mater. 2010, 33, 236–240.

83. Prudenzano, F.; Mescia, L.; Palmisano, T.; Surico, M.; De Sario, M.; Righini, G.C. Optimization of pump absorption in MOF lasers via multi-long-period gratings: Design strategies. Appl. Opt. 2012, 51, 1410–1420.

84. Slavík, R.; Kulishov, M.; Park, Y.; Azaña, J. Long-period fiber-grating-based filter configuration enabling arbitrary linear filtering characteristics. Opt. Lett. 2009, 34, 1045–1047.

85. Kim, S.J.; Eom, T.J.; Lee, B.H.; Park, C.S. Optical temporal encoding/decoding of short pulses using cascaded long period fiber gratings. Opt. Express 2003, 11, 3034–3040.

86. Ashrafi, R.; Li, M.; Azaña, J. Tsymbol/s optical coding based on long-period gratings. IEEE Photonics Technol. Lett. 2013, 25, 910–913.

87. Slavik, R.; Park, Y.; Kulishov, M.; Azaña, J. Terahertz-bandwidth high-order temporal differentiators based on phase-shifted long-period fiber gratings. Opt. Lett. 2009, 34, 3116–3118.

88. Ashrafi, R.; Li, M.; Azaña, J. Coupling-strength-independent long-period grating designs for THz-bandwidth optical differentiators. IEEE Photonics J. 2013, 5, 7100311.

89. Shu, X.; Allsop, T.; Gwandu, B.; Zhang, L.; Bennion, I. High-temperature sensitivity of long-period gratings in B-Ge codoped fiber. IEEE Photonics Technol. Lett. 2001, 13, 818–820.

90. Wang, Z.; Ramachandran, S. Ultrasensitive long-period fiber gratings for broadband modulators and sensors. Opt. Lett. 2003, 28, 2458–2460.

91. Steinvurzel, P.; Moore, E.D.; Mägi, E.C.; Eggleton, B.J. Tuning properties of long period gratings in photonic bandgap fibers. Opt. Lett. 2006, 31, 2103–2105.

92. Potyrailo, R.A.; Hobbs, S.E.; Hieftje, G.M. Optical waveguide sensors in analytical chemistry: Today's instrumentation, applications and trends for future development. Fresenius J. Anal.Chem. 1998, 362, 349–373.

93. McAdam, G.; Newman, P.J.; McKenzie, I.; Davis, C.; Hinton, B.R.W. Fiber optic sensors for detection of corrosion within Aircraft. Struct. Health Monit. 2005, 4, 47–56.

94. Mendoza, E.A.; Khalil, A.N.; Sun, Z.; Robinson, D.; Syracuse, S.J.; Egalon, C.O.; Gunther, M.F.; Lieberman, R.A. Embeddable distributed moisture and pH sensors for nondestructive inspection of aircraft lap joints. Proc. SPIE 1995, 2455, 102–112.

95. Mendoza, E.A.; Robinson, D.; Lieberman, R.A. Miniaturized integrated optic chemical sensors for environmental monitoring and remediation. Proc. SPIE 1996, 2836, 76–87.

96. Cordero, S.R.; Beshay, M.; Low, A.; Mukamal, H.; Ruiz, D.; Lieberman, R.A. A distributed fiber optic chemical sensor for hydrogen cyanide detection. Proc. SPIE 2005, 5993, 599302.

97. Cordero, S.R.; Mukamal, H.; Low, A.; Locke, E.P.; Lieberman, R.A. A fiber optic sensor for nerve agent. Proc. SPIE 2006, 6378, 63780U–63783U.

98. Mukamal, H.; Cordero, S.R.; Ruiz, D.; Beshay, M.; Lieberman, R.A. Distributed fiber optic chemical sensor for hydrogen sulfide and chlorine detection. Proc. SPIE 2005, 6004, 600406.

99. Ghandehari, M.; Vimer, C.S. In situ monitoring of pH level with fiber optic evanescent field spectroscopy. NDT E Int. 2004, 37, 611–616.

100. Sinchenko, E.; Gibbs, W.E.K.; Mazzolini, A.P.; Stoddart, P.R. The effect of the cladding refractive index on an optical fiber evanescent-wave sensor. J. Lightwave Technol. 2013, 31, 3251–3257.

101. Stewart, G.; Culshaw, B. Optical waveguide modelling and design for evanescent field chemical sensors. Opt. Quantum Electron. 1994, 26, S249–S259.

102. Schwotzer, G.; Latka, I.; Lehmann, H.; Willsch, R. Optical sensing of hydrocarbons in air or in water using UV absorption in the evanescnt field of fibers. Sens. Actuators B 1997, 38–39, 150–153.

103. Gupta, B.D.; Dodeja, H.; Tomar, A.K. Fibre-optic evanescent field absorption sensor based on a U-shaped probe. Opt. Quantum Electron. 1996, 28, 1629–1639.

104. Zimmerman, B.; Burck, J.; Ache, H.J. Studies on siloxane polymers for NIR-evanescent wave absorbance sensor. Sens. Actuators B 1997, 41, 45–54.

105. Qing, D.-K.; Yamaguchi, I. Analysis of the sensitivity of optical waveguide chemical sensor for TM modes by the group-index method. J. Opt. Soc. Am. B 1999, 16, 1359–1369.

106. Buerck, J.; Roth, S.; Kraemer, K.; Scholz, M.; Klaas, N. Application of a fiber-optic NIR-EFA sensor system for in situ monitoring of aromatic hydrocarbons in contaminated groundwater. J. Hazard. Mater. 2001, 83, 11–28.

107. Prudenzano, F.; Mescia, L.; Allegretti, L.A.; Calò, G.; D'Orazio, A.; De Sario, M.; Palmisano, T.; Petruzzelli, V. Design of an optical sensor array for hydrocarbon monitoring. Opt. Quantum Electron. 2009, 41, 55–68.

108. Mescia, L.; Prudenzano, F.; Allegretti, L.; Calò, G.; De Sario, M.; D'Orazio, A.; Maiorano, L.; Palmisano, T.; Petruzzelli, V. Design of silica-based photonic crystal fiber for biosensing applications. J. Non-Cryst. Solids 2009, 355, 1163–1166.

109. Palmisano, T.; Prudenzano, F.; Warren-Smith, S.C.; Monro, T.M. Design of exposed-core fiber for methadone monitoring in biological fluids. J. Non-Cryst. Solids 2011, 357, 2000–2004.

110. Kostecki, R.; Ebendorff-Heidepriem, H.; Davis, C.; McAdam, G.; Warren-Smith, S.C.; Monro, T.M. Silica exposed-core microstructured optical fibers. Opt. Mater. Express 2012, 2, 1538–1547.

111. Eggleton, B.J.; Kerbage, C.; Westbrook, P.S.; Windeler, R.S.; Hale, A. Microstructured optical fiber devices. Opt. Express 2001, 9, 698–713.

112. Bise, R.T.; Windeler, R.S.; Kranz, K.S.; Kerbage, C.; Eggleton, B.J.; Trevor., D.J. Tunable photonic band gap fiber. In Proc. Opt. Fiber Commun. Conf. Exhib. OSA Trends Opt. Photon. (Opt. Soc.Am. Washington); 2002; Volume 70, pp. 466–468.

113. Wang, Y.; Tan, X.; Jin, W.; Ying, D.; Hoo, Y.L.; Liu, S. Temperature controlled transformation in fiber types of fluid-filled photonic crystal fibers and applications. Opt. Lett. 2010, 35, 88–90.

114. Yiou, S.; Delaye, P.; Rouvie, A.; Chinaud, J.; Frey, R.; Roosen, G.; Viale, P.; Février, S.; Roy, P.; Auguste, J.-L.; Blondy, J.-M. Stimulated Raman scattering in an ethanol core microstructured optical fiber. Opt. Express 2005, 13, 4786–4791.

115. Huang, Y.; Xu, Y.; Yariv, A. Fabrication of functional microstructured optical fibers through a selective-filling technique. Appl. Phys. Lett. 2004, 85, 5182–5184.

116. Cordeiro, C.M.B.; dos Santos, E.M.; Cruz, C.H.B.; de Matos, C.J.S.; Ferreira, D.S. Lateral access to the holes of photonic crystal fibers—Selective filling and sensing applications. Opt. Express 2006, 14, 8403–8412.

117. Meneghini, C.; Caron, S.; Poulin, A.C.J.; Proulx, A.; Émond, V.; Paradis, P.; Paré, C.; Fougères, A. Determination of ethanol concentration by Raman spectroscopy in liquid-core microstructured optical fiber. IEEE Sens. J. 2008, 8, 1250–1255.

118. Ren, H.C.; Vollmer, F.; Arnold, S.; Libchaber, A. High-Q microsphere biosensor-analysis for adsorption ofrodlike bacteria. Opt. Express 2007, 15, 17410–17423.

119. Vollomer, F.; Stephen, A. Whispering-gallery-mode biosensing: Label-free detection down to single molecules. Nat. Methods 2008, 5, 591–596.

120. Yang, L.; Vahala, K.J. Gain functionalization of silica microresonators. Opt. Lett. 2008, 28, 592–594.

121. Murugan, G.S.; Zervas, M.N.; Panitchob, Y.; Wilkinson, J.S. Integrated Nd-doped borosilicate glass microsphere laser. Opt. Lett. 2011, 36, 73–75.

122. Chen, S.Y.; Sun, T.; Grattan, K.T.V.; Annapurna, K.; Sen, R. Characteristics of Er and ErYbCr doped phosphate microsphere fibre lasers. Opt. Commun. 2009, 282, 3765–3769.

123. Feron, P. Whispering Gallery Mode Lasers in Erbium doped fluoride glasses. Annales de la Fondation Louis de Broglie 2004, 29, 317–329.

124. Nunzi Conti, G.; Chiasera, A.; Ghisa, L.; Berneschi, S.; Brenci, M.; Dumeige, Y.; Pelli, S.; Sebastiani, S.; Feron, P.; Ferrari, M.; Righini,

G.C. Spectroscopic and lasing properties of Er3+ doped glass microspheres. J. Non-Cryst. Solids 2006, 352, 2360–2363.

125. Elliot, G.R.; Hewak, D.W.; Senthil Murugan, G.; Wilkinson, J.S. Chalcogenide glass microspheres; their production, characterization and potential. Opt. Express 2007, 15, 17542–17553.

126. Mescia, L.; Prudenzano, F.; De Sario, M.; Palmisano, T.; Ferrari, M.; Righini, G.C. Design of rare-earth-doped microspheres. IEEE Photonics Technol. Lett. 2010, 22, 422–424.

127. Kouki, T.; Makoto, T. Optical microsphere amplification system. Opt. Lett. 2007, 32, 3197–3199.

128. Boucher, Y.G.; Feron, P. Generalized transfer function: A simple model applied to active single-mode microring resonators. Opt. Commun. 2009, 282, 3940–3947.

129. Mescia, L.; Bia, P.; De Sario, M.; Di Tommaso, A.; Prudenzano, F. Design of mid-infrared amplifiers based on fiber taper coupling to erbium-doped microspherical resonator. Opt. Express 2012, 20, 7616–7629.

130. Mescia, L.; Bia, P.; Losito, O.; Prudenzano, F. Design of Mid-IR Er3+-doped microsphere laser. IEEE Photonics J. 2013, 5, 1501308.

CHAPTER 4

The Application of Optical CDMA-Based Fiber Radio Networks in Wireless Sensor Networks

Chao-Chin YANG

Department of Electronic Engineering, Kun Shan University, Tainan, Taiwan, China

ABSTRACT

One fiber radio scheme using shifted prime codes for interference elimination is proposed for optical code-division multiple-access (OCDMA) network. By taking advantage of the cyclic property of the shifted prime codes in the same code groups, the proposed compact decoder is low cost and suitable to be used in the task manager node in the applications of wireless sensor networks. The performance comparison for several OCDMA-based fiber radio networks is also given to clarify the advantage of the proposed one.

KEYWORDS

Optical Code-Division Multiple-Access (OCDMA), Passive Optical Network (PON), Arrayed Waveguide Grating (AWG), Prime Code

1. INTRODUCTION

In today's mobile radio networks, the provision of broadband services between a large number of remote base stations (RBSs) is required. To develop simple and low-cost RBSs for high rate transmission, the combination of mobile radio and optical fiber network is one possible solution since the components for complicated operation of signal processing can be moved to control base station (CBS) [1]. In the case of wireless sensor networks (WSNs) nowadays, the condition may be similar [2]. The sensor nodes need to route the collected data back to the sinks, and the sinks may communicate with the task manager node via networks for long distance transmission. Therefore, fiber radio networks can also provide the communication between the sinks (e.g. RBS) and the task manager node (e.g. CBS) to simplify the design of the sinks.

Code division multiple access (CDMA) techniques were investigated for optical network applications during the last 20 to 25 years. These techniques allow many users to access the common channel asynchronously and securely, and, since dedicated time or wavelength slots do not have to be allocated, high statistical multiplexing gain can be offered even in bursty traffic. These characteristics distinguish CDMA from other multiplexing schemes such as time division multiple access (TDMA) and wavelength division multiple access (WDMA). Several OCDMA-based fiber radio schemes were proposed [3-6]. Among these, fiber radio schemes based on spectral-amplitude-coding (SAC) had the advantages of multiple access interference (MAI) elimination and the avoidance of sampling before the optical encoding process [4-6].

A code that can be used in the SAC scheme is denoted as (N, w, λ), where N is code length, w is code weight, and λ is the in-phase cross correlation [6]. Originally unipolar m-sequences were used in the SAC-based networks [7], and, to enhance the network performance, code sequences with complementary code keying ability such as Hadamard codes [7] could be used. However, since these two code families had code weights equivalent to half of the corresponding codelengths, they induced serious phase-induced intensity noise (PIIN) in the photodiodes of the decoders. To suppress the phase-induced intensity noise in the photodiodes of the decoders, several codes with ideal λ (λ = 1) were proposed, such as Balanced Incomplete Block Design (BIBD) codes [8]. The former had fully cyclic property and N × N arrayed waveguide grating (AWG) can be used as compact coding devices [9,10]. However, when the total number of users in the network increased, a large number of AWG ports were required and the effect of crosstalk was more serious.

To enlarge the network capacity further, new coding schemes using codewords with small λ and having the two-code keying ability can be developed. In [11], one coding scheme was proposed and code sequences with λ=0 or 1 can be used for two-code keying of information bits. By the use of the cyclic property for shifted prime (SP) codes, compact encoder structure based on cascaded AWGs and simplified decoder based on fiber Bragg grating (FBG) were also proposed for the application of passive optical networks (PONs). The optical line terminal (OLT) encoder based on cascaded AWGs had the advantage that only p×p and 1×p AWGs were needed in the encoder for SP codes with codelength p^2.

In this paper, the application of SAC schemes using SP codes in radio fiber network is explored. By the modification of the encoder/decoder pair for the downstream transmission in the PON in [11], the encoder/decoder implementation in the upstream transmission (from RBSs to CBS) for the fiber radio network can be obtained. Performance analysis takes thermal noise and PIIN into account. It shows that, as compared to several fiber radio schemes based merely on SAC, the use of SP codes allows larger number of RBSs to access the network simultaneously under a given carrier-to-noise ratio (CNR).

2. FIBER RADIO NETWORK BASED ON SP CODE

The proposed SAC OCDMA-based fiber radio network consists of one CBS, the RBSs connected , and one 1×M splitter connected to the CBS and RBSs by optical fibers, where M is the total number of RBSs in the network. Each RBS is assigned two SP codewords for optical transmission. The RBSs receive the radio signals from the mobile terminals or wireless sensor networks, and, after simply down-conversion process, encode these radio signals with the assigned SP codewords. The resulting encoded optical signals are combined in the 1 × M coupler and transmitted to the CBS through one fiber. Finally the decoders in the CBS receive the signals from the 1 × M coupler and extract the radio signal of each RBS for further processing.

Table 1. The SP codewords for p=3.

e	f	$\mathbf{C}_{e,f}$	$\mathbf{X}_{e,f}$
0	0	0 0 0	100 100 100
0	1	1 1 1	010 010 010
0	2	2 2 2	001 001 001
1	0	0 1 2	100 010 001
1	1	1 2 0	010 001 100
1	2	2 0 1	001 100 010
2	0	0 2 1	100 001 010
2	1	1 0 2	010 100 001
2	2	2 1 0	001 010 100
3	0		111 000 000
3	1		000 111 000
3	2		000 000 111

In the following, the property of SP codes is briefly reviewed: One SP code $X_{e,f}$'s with length L = p^2 used in this paper can be obtained from one modified stuffed shifted prime (MSSP) code with the same parameter p in [12] by eliminating the extra codeword and the last p+1 chips of each remaining MSSP codewords. For example, Table 1 shows the SP codewords for p = 3, and ,from [12], the cross-correlation between $X_{e,f}$ and $X_{e',f'}$ is

$$X_{e,f} \odot X_{e',f'} = \begin{cases} p, e=e', f=f', \\ 0, e=e', f \neq f', \\ 1, \ e \neq e', \end{cases} \tag{1}$$

and thus

$$X_{e,f_1} \odot X_{e',f'} - X_{e,f_2} \odot X_{e',f'} = \begin{cases} p, \ e=e', f_1=f', \\ -p, \ e=e', f_2=f', \\ 0, \ \text{otherwise.} \end{cases} \tag{2}$$

where $\odot$ is the dot-product of two vectors. Since the RBSs can use the code property in Equation (2) for MAI elimination, any two distinct SP codewords with the same value of e can be assigned to one specific RBS for the encoding of radio signals, and the value of M is (p+1)(p-1)/2. However, the two codewords used for decoding in conventional SAC schemes are always complement to each other. Therefore, this coding scheme has the advantage of security enhancement in the SAC-based fiber radio networks since the eavesdroppers need to guess two codewords used for encoding.

3. IMPLEMENTATION OF CODER/DECODER

The RBSs of the proposed fiber radio network use the encoder shown in Figure 2, which is designed for the first SP codewords X_{e,f_1} = [1 0 0 1 0 0 1 0 0] and the second SP X_{e,f_2} = [0 1 0 0 1 0 0 1 0] of this RBS. Light from the broadband light source is incident to one FBG-based encoder cascading with another FBG-based encoder. Suppose that each wavelength chip has spectral width $\Delta\lambda$, the 3dB bandwidths of the FBGs in these two FBG-based encoders should also be $\Delta\lambda$. The center wavelengths of the FBGs in these two FBG-based encoders are { λ_0, λ_3, λ_6}(corresponding to X_{e,f_1}) and { λ_1, λ_4, λ_7 }(corresponding to X_{e,f_2}), respectively. Therefore, the first (or second) codeword of this RBS is generated in the input port of the upper (or lower) intensity modulator (IM) and is intensity-modulated by the signal (1+ d(t))/2 (or (1- d(t))/2). Here d(t) is the radio signal normalized from the radio signal d'(t) such that max |d(t)| = 1. To make the receiver bandwidth B as small as possible, the frequency of the received radio signal from the antenna can be first converted to lower frequency before the signal manipulation process.

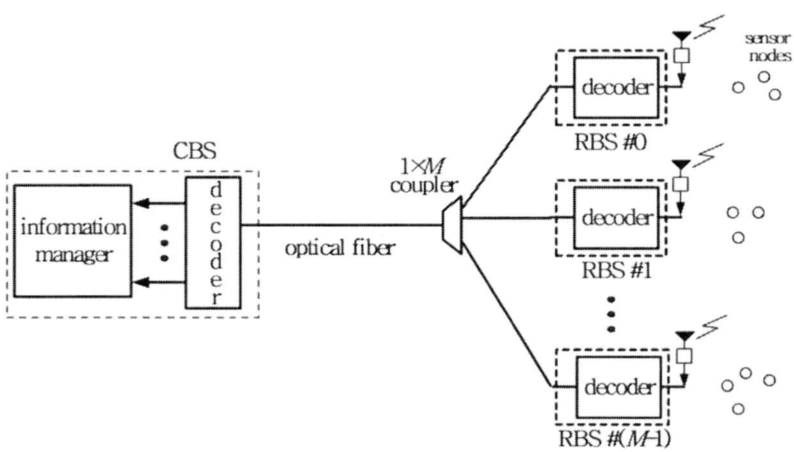

Figure 1. The proposed fiber radio network.

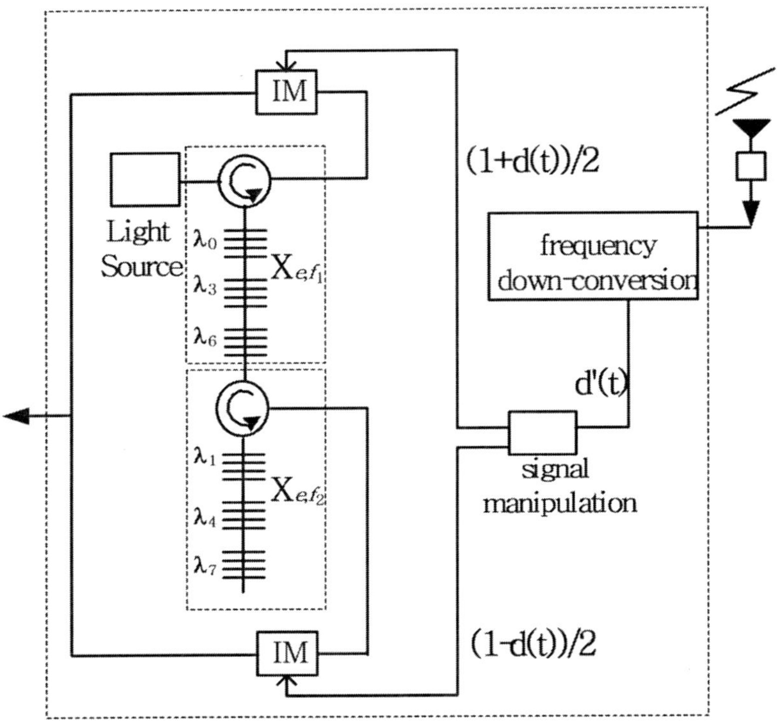

Figure 2. The encoder in the RBS.

The decoder used to decode all the SP codewords from the RBSs with p = 3 is shown in Figure 3, which is similar to the AWG-based encoder in [11]. Two kinds of AWGs are used in this encoder: One is coarse AWG with free spectral range $FSR_c = p^2 \Delta\lambda$ and channels bandwidth $p\Delta\lambda$, and the other is fine AWG with $FSR_f = p\Delta\lambda$ and channels bandwidth $\Delta\lambda$ [9]. Note that the center wavelengths of the first output port at the coarse AWG are $\lambda_c - FSR_f \pm r\, FSR_c$ (r =0,1,...), thus there is only one coarse AWG channel at each coarse AWG output port within the encoded spectrum. However, for the fine AWG, the center wavelengths of the AWG channels that route from the first input port to the first output port are $\lambda_0 = \lambda_c - (p-1)\,\Delta\lambda/2 \pm sFSR_f$ (s = 0,1,...), thus there are p wavelength channels within the encoded spectrum that communicates between any input-output port pair.

The general principle to construct the decoder for one SP code family $X_{e,f}$'s with code length L=p^2 is described as follow: the single light source is connected to one 1× (p + 1) splitter. The output port # e (e = 0 ~ p - 1) of the splitter is connected to one coarse AWG cascaded with one fine AWG and the output port #i of the coarse AWG is connected to the input port # $C_{e,0}$ (i) of the fine AWG. Take the encoder (p = 3) in Figure 1(a) as example, light emitted from the filtered light source contains wavelength components$\lambda_0 \sim \lambda_8$ ($\Delta\lambda=\lambda_i-\lambda_{i+1}$) and is split into four. At these four encoder branches, wavelength chips λ_{0+bp}, λ_{1+bp} and λ_{2+bp} are demultiplexed and appear at the output port # b of the coarse AWGs. Due to the connections between the coarse and fine AWG ports described above, $\lambda_{f\oplus 0}$, $\lambda_{p+(f\oplus 1)}$ and $\lambda_{2p+(f\oplus 1)}$ (f = 0, 1, ...,p-1) corresponding to $X_{0,f}$ appears at the fine AWG output port # f at encoder branch #0 and $\lambda_{f\oplus 1}$, $\lambda_{p+(f\oplus 0)}$ and $\lambda_{2p+(f\oplus 1)}$ corresponding to $X_{1,f}$ appears at the fine AWG output port # f at encoder branch #1($\oplus$ denotes the modulo-p addition). Let's take the generation of $X_{1,0}$ as example. Wavelengths λ_0, λ_3 and λ_6 from the light source are demultiplexed by the coarse AWG at encoder branch #1 and they appear at the output port #0, #1 and #2, respectively. The output ports # 0, #1 and #2 of this coarse AWG are connected to the input ports #0, #1 and #2 of one fine AWG according to $C_{1,0}$ (= [0 1 2]), respectively, and thus λ_0, λ_3 and λ_6 appear at the input ports #0, #1 and #2 of this fine AWG, respectively. Since wavelength λ_g from the input port # i appears at output port # (g$\oplus$(-i)) for each fine AWG [9], λ_0, λ_3 and λ_6 all appears at output port #0 of this fine AWG which corresponds to $X_{1,0}$ = [100100100].

In this way, the first p^2 codewords can be obtained at the fine AWG output ports of the encoder. Since the wavelengths appeared at the coarse AWG output port # f of the last encoder branch correspond to $X_{3,f}$, the remaining p additional codewords in group #3 are obtained. Balanced detection for each RBS's signal can be accomplished with one balanced detector shown in Figure 3. The upper and lower photo-diodes of the balanced detector for one RBS are connected to the fine AWG output ports in the decoder according to the first and second codewords assigned to that RBS, respectively. Thus each balanced photo-detector can generate electrical current proportional to the value of $X_{e,f_1} \odot S - X_{e,f_2} \odot S$ and the MAI elimination scheme in Equation (2) is realized, where S is the summation signals from all RBSs.

4. PERFORMANCE ANALYSIS

Assume the light source in the RBS is unpolarized and has flat spectrum for encoding with bandwidth $\Delta\nu$ and magnitude Psr/ $\Delta\nu$, where Psr is the effective power received by the photo-diodes in the decoder of the CBS. The average carrier power in the balanced detector is $P_c = R^2 P_{sr}^2 /(2p^2)$, where R is the responsivity of the photodiodes. For the fiber radio network adopting SP codes, the PSD of the upstream optical signal arrived at the input port of the decoder in the CBS can be written as [5,10]

$$s(\nu)=\frac{P_{sr}}{\Delta\nu}\sum_{k=0}^{K-1}\{\frac{1}{2}(1+d_k(t))\sum_{i=0}^{L-1}x_{k,i}^{(0)}\Pi(\nu,i)+\frac{1}{2}(1-d_k(t))\sum_{i=0}^{L-1}x_{k,i}^{(1)}\Pi(\nu,i)\} \tag{3}$$

Where

$$\Pi(\nu,i)=\begin{cases}1, i\le L(\frac{\nu-\nu_0}{\Delta\nu}+\frac{1}{2})\le i+1\\ 0,\ \text{otherwise.}\end{cases} \tag{4}$$

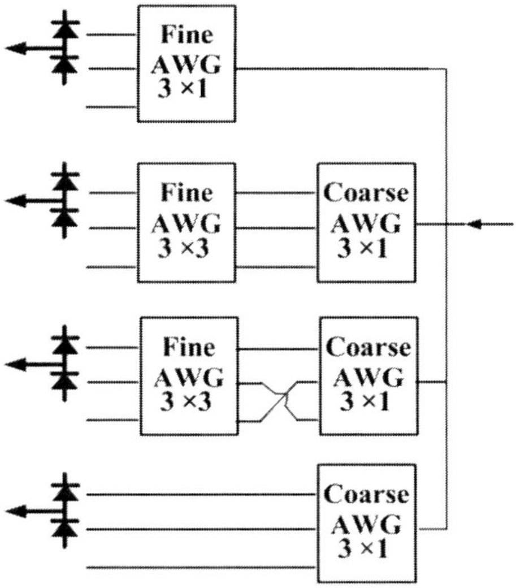

Figure 3. The decoder for SP codes in the CBS.

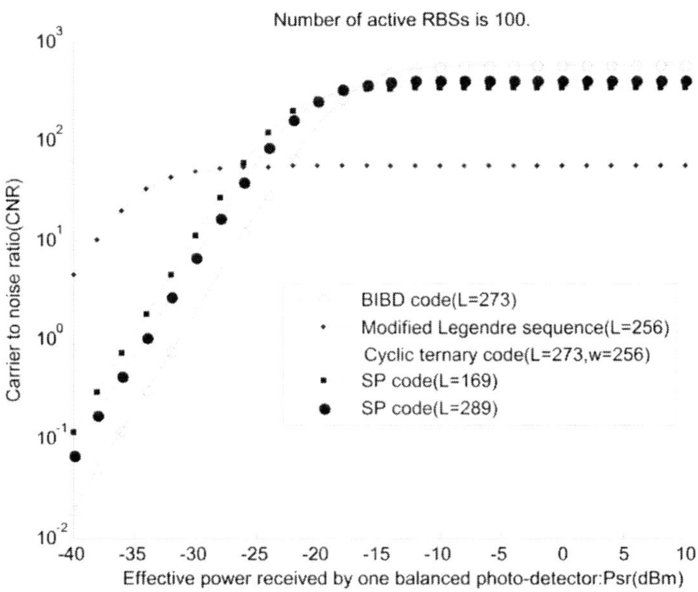

Figure 4. CNR vs. Psr.

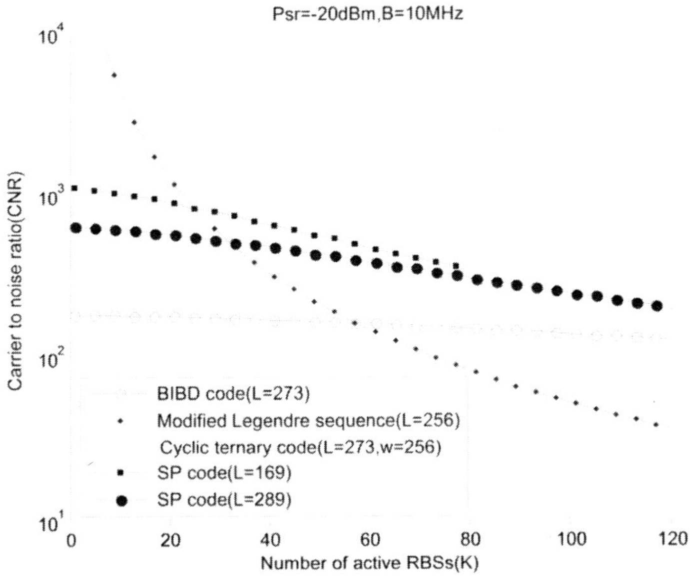

Figure 5. CNR vs. K.

Here K is the number of active RBSs and $d_k(t)$ is the normalized radio signal of k-th RBS. To simplify the notation, the i-th chip of the first and second codewords of k-th RBS are denoted as $c_{k,i}^{(0)}$ and $c_{k,i}^{(1)}$, respectively. Assume the sensor nodes use modulation with constant envelope such as phase shift keying. If at this time only one sensor node transmit signal to the 0-th RBS, the radio signal is approximately a sine wave and the assumption of time average amplitude $< d_0(t) > = 0$ and time average power $< d_0(t)^2 > = 0.5$ can be used [5,6]. Thus the power of phase-induced intensity noise (PIIN) can be obtained [5,10]:

$$< I_{\text{PIIN}}{}^2 >= BR^2 \left[\int_0^\infty G_0^2(v)\,dv + \int_0^\infty G_1^2(v)\,dv \right]$$
$$= \frac{BR^2}{p^2 \Delta v} P_{sr}{}^2 \left(\frac{3}{4}p + \frac{7}{2}K_{eff} + \frac{2}{p}K_{eff}(K_{eff}-1)\right) \quad (5)$$

where B is noise-equivalent electrical bandwidth of the receiver. The effective number of interfering active RBS is [11]

$$K_{eff} = K - 1 - \left\lfloor \frac{K-1}{p} \right\rfloor \tag{6}$$

For wireless sensor networks using the 915MHz industrial, scientific, and medical (ISM) band for radio frequency transmission, the data rate may be just about 10k Hz. [1] Therefore, the carrier frequency of the radio signals received by the RBSs can be converted to about 10MHz by the frequency down-conversion circuit in the Figure 2, and B is set to 10MHz here. Other parameters used in the performance analysis are R = 0.8A/W, Δv = 7.5 THz, and v_0 = 193.1THz and the power of thermal noise is $\langle i_{th}^2 \rangle = 16*10^{-17} A^2$.</i

The relationship between CNR and K are shown in Figure 4. Here the results for other three codes (e.g. balanced incomplete block design (BIBD) codes, modified Legendre sequences mentioned in [6] and cyclic ternary sequences with weight w =256 in [6] are also shown for comparison and their code lengths are approximately the same. (Note that Hadamard codes in [5,7] obtain the same performance as modified Legendre sequences, and thus only the results for modified Legendre sequences are shown here.) When Psr is low, the influence of PIIN is less and the CNR for each code increases with Psr. This increase gradually disappears due to the domination of PIIN. The CNR curves for modified Legendre sequences and cyclic ternary sequences are approximately the same and these two codes obtain better CNRs then BIBD and SP codes when Psr is low. However, when Psr is high, the situation is reversed. Though the BIBD code obtains slightly better CNRs than SP code with similar length when Psr is larger than -15dBm, it requires very high transmitted power from the light sources in the RBSs when the insertion losses in the encoder/ decoder are taken into consideration. Since the insertion losses of FBG-based encoder in one RBS and the

AWG-based decoder in the CBS are approximately 30dB [11], the CNR performance for Psr approximately -20dBm seems more important when the transmitted power from the light source in the RBS is about 10dBm [1], and SP code with L=289 obtains better CNR at Psr = -20dBm for K = 100.

The relationship between CNR and K for the codes in Figure 4 is shown in Figure 5, where Psr is set at -20dBm. It is found that all four codes obtain lower CNR when K increases. When K is relatively small, modified Legendre sequences and cyclic ternary sequences obtain better CNRs then BIBD and SP codes. However, the situation is reversed when K is relatively large. Though SP codes obtain better CNRs than BIBD codes for Psr = -20dBm, the corresponding cardinality is about half of that for BIBD codes with similar code lengths. Since SP codes can use AWG routers with smaller sizes for decoder implementation in the CBS as compared to BIBD codes, it is necessary to consider the tradeoffs between these codes for better network design.

5. CONCLUSIONS

One fiber radio scheme for the application of wireless sensor networks is proposed for optical code-division multiple-access (OCDMA) network. By the use of this fiber radio scheme, the design of sinks in the wireless sensor networks can be simplified, which has advantages for temporary deployment of the sensor networks. In addition, the proposed decoder in the CBS is compact and these features make the proposed fiber radio scheme more attractive for further research.

REFERENCES

1. H. Kim, J. M. Cheong, C. H. Lee, and Y. C. Chung, "Passive optical network for microcellular CDMA personal communication service,"

IEEE Photonics Technology Letters, vol. 10, no. 11, pp. 1641–1643, November 1998.

2. F. Akyildiz, W. Su, Y. Sankarasubmmaiam, and E. Cayirci, "A Survey on sensor networks," IEEE Communication Magazine, vol. 91, no. 8, pp. 102–114, August 2002.
3. K. Tsukamoto, T. Higashino, T. Nakanishi, and S. Komaki, "Direct optical switching code-division multiple-access system for fiber-optic radio highway networks," IEEE Journal of Lightwave Technology, vol. 21, no. 12, pp. 3209–3220, December. 2003.
4. B. K. Kim, S. Park, Y. Yeon, and B. W. Kim, "Radio-over-fiber system using fiber-grating-based optical CDMA with modified PN codes," IEEE Photonics Technology Letters, vol. 15, no. 10, pp. 1485–1487, October 2003.
5. T. Demeechai, "Noise-limited performance of spectral-amplitude-coding optical CDMA in fibre-optic radio highway networks," IEE Proceedings of Optoelectronics, vol. 152, no. 5, pp. 269–273, October 2005.
6. C. C. Yang, "Optical CDMA fiber radio networks using cyclic ternary sequences," IEEE Communications Letters, vol. 12, no. 1, pp. 41–43, January 2008.
7. M. Kavehrad and D. Zaccarin, "Optical code-division-
8. multiplexed systems based on spectral encoding of noncoherent sources," IEEE Journal of Lightwave Technology, vol. 13, no. 3, pp. 534–545, March 1995.
9. X. Zhou, H. M. H. Shalaby, C. Lu, and T. Cheng, "Code for spectral amplitude coding optical CDMA systems," Electronics Letters, vol. 36, pp. 728–729, April 13, 2000.
10. B. Glance, I. P. Kaminow, and R. W. Wilson, "Applications of the integrated waveguide grating router," IEEE/OSA Journal of Lightwave Technology, vol. 12, pp. 957–962, June 1994.
11. C. C. Yang, "The application of spectral-amplitude-codi-
12. ng optical CDMA in passive optical networks," Optical Fiber Technology, vol. 14, no. 2. pp. 134–142, April 2008.

13. C. C. Yang, "Optical CDMA passive optical network using prime code with interference elimination," IEEE Photonics Technology Letters, vol. 19, no. 7, pp. 516–518, April 2007.
14. C. C. Yang, J. F. Huang, and T. C. Hsu, "Differentiated service provision in optical CDMA network using power control," IEEE Photonics Technology Letters, vol. 20, no. 20, pp. 1664–1666, October 2008.

CHAPTER 5

Recent Development in Optical Fiber Biosensors

María Espinosa Bosch [1], Antonio Jesús Ruiz Sánchez [2], Fuensanta Sánchez Rojas [3,*] and Catalina Bosch Ojeda [3]

[1] Department of Pharmacy, General Hospital, University Hospital Virgen del Rocío, Manuel Siurot s/n, 41013, Sevilla, Spain

[2] Department of Organic Chemistry, Faculty of Sciences, University of Málaga, Campus Teatinos s/n, 29071, Málaga, Spain

[3] Department of Analytical Chemistry, Faculty of Sciences, University of Málaga, Campus Teatinos s/n, 29071, Málaga, Spain

ABSTRACT

Remarkable developments can be seen in the field of optical fibre biosensors in the last decade. More sensors for specific analytes have been reported, novel sensing chemistries or transduction principles have been introduced, and applications in various analytical fields have been realised. This review consists of papers mainly reported in the last decade and presents about applications of optical fiber biosensors.

Discussions on the trends in optical fiber biosensor applications in real samples are enumerated.

KEYWORDS

Optical fiber; biosensor; applications; review

1. INTRODUCTION

Biosensor development is driven by the continuous need for simple, rapid, and continuous in-situ monitoring techniques in a broad range of areas, e.g. medical, pharmaceutical, environmental, defense, bioprocessing, or food technology.

Biosensors make use of biological components in order to sense a species of interest (which by itself need not be a "biospecies"). On the other side, chemical sensors not using a biological component but placed in a biological matrix are not biosensors by definition. Biological systems (such as tissues, micro-organisms, enzymes, antibodies, nucleic acids, etc.) when combined with a physico-chemical transducer (optical, electrochemical, thermometric, piezoelectric) form a biosensor.

On the other hand, the development of optical-fiber sensors during recent years is related to two of the most important scientific advances: the laser and modern low-cost optical fibers. Recently, optical fibers have become an important part of sensor technology. Their use as a probe or as a sensing element is increasing in clinical, pharmaceutical, industrial and military applications. Excellent light delivery, long interaction length, low cost and ability not only to excite the target molecules but also to capture the emitted light from the targets are the main points in favour of the use of optical fibers in biosensors.

Optical fibers transmit light on the basis of the principle of total internal reflection (TIR). Fiber optic biosensors are analytical devices in which a fiber optic device serves as a transduction element. The usual aim is to produce a signal that is proportional to the concentration of a chemical o biochemical to which the biological element reacts. Fiber optic biosensors are based on the transmission of light along silica glass fiber, or plastic optical fiber to the site of analysis. Optical fiber biosensors can be used in combination with different types of spectroscopic technique, e.g. absorption, fluorescence, phosphorescence, surface plasmon resonance (SPR), etc.

Optical biosensors based on the use of fiber optics can be classified into two different categories: intrinsic sensors, where interaction with the analyte occurs within an element of the optical fiber; and extrinsic sensors, in which the optical fiber is used to couple light, usually to and from the region where the light beam is influenced by the measurand.

Biosensors are attractive because they can be easily used by non-specialist personnel and they allow accurate determination with either no or minimal sample treatment. Therefore, fiber optic biosensors may be especially useful in routine tests, patient home care, surgery and intensive care, as well as emergency situations.

2. ABSORBANCE MEASUREMENTS

The simplest optical biosensors use absorbance measurements to determine any changes in the concentration of analytes that absorb a given wavelength of light. The system works by transmitting light through an optical fiber to the sample; the amount of light absorbed by the analyte is detected through the same fiber or a second fiber. The biological material is immobilized at the distal end of the optical fibers and either produces or extracts the analyte that absorbs the light.

A fiber optic pH sensor [1] and a fiber optic oxygen sensor [2] have been developed by Wolthuis et al., for use in medical applications. In the first case, the sensor uses an absorptive indicator compound with a long wavelength absorption peak near 625 nm; change in absorption over the pH range 6.8 to 7.8 is reasonably linear. The sensor is interrogated by a pulsed, red LED. Return light signal is split into short and long wavelength components with a dichroic mirror; the respective signals are detected by photodiodes, and their photocurrents are used to form a ratiometric output signal. In laboratory tests, the sensor system provided resolution of 0.01 pH and response time of 30-40 s. Following gamma sterilization, laboratory sensor testing with heparinised human blood yielded excellent agreement with a clinical blood gas analyzer. Excellent sensor performance and low cost, solid-state instrumentation are hallmarks of this sensor-system design. In the second, the sensor's viologen indicator becomes strongly absorbant after brief UV stimulation, and then returns to the transparent state. The rate of indicator return to transparency is proportional to the local oxygen concentration. Indicator absorbance is monitored with a red LED. The solid state sensor system has performance comparable to existing oxygen measurement techniques, and may be applicable for both in vitro and in vivo oxygen measurements.

Molecular recognition processes are very important in the investigation of biological phenomena as well as development of fiber optic biosensor. As promising materials, biofunctionalized polydiacetylene lipids have many special properties. p-10,12-pentacosadiyne-1-N-(3,6,9-trioxaundecylanide) -α-D-mannopyranoside (MPDA) was synthesized from 10,12-pentacasadiynoic acid (PDA). The mixed monolayer MPDA/PDA was prepared by using Langmuir-Blodgett technique on glass and optical fiber, respectively. Molecular recognition of Escherichia coli (E. coli) resulted in a change of the film colour and can be quantified by UV-VIS absorption spectroscopy. The properties might be used for screening drugs and diagnosis [3].

On the other hand, thin films of polyaniline (PANI) and polypyrrole (PPy) have been shown to respond to pH and to the redox state of a solution, thereby undergoing spectral changes in the visible and near infrared. They also may serve as a matrix for enzyme immobilization. PANI films are easily prepared, and spectral changes depending on analyte concentration occur at wavelengths at which low cost lasers and LEDs are available. PANI and PPy are conductive organic polymers that can be driven between different oxidation states by chemical or electrochemical methods. PANI has two redox equilibriums associated with the polymer chain, and the optical properties of the polymer are functions of the state of protonation and the oxidation state. Since protons and electrons are directly involved in the polymer redox reaction, changes in optical spectra (e.g. absorbance) can be attributed to the concentration of protons or the number of electrons produced in an enzymatic reaction. PANI films of good optical quality can be produced by chemical means on almost any desired support including polystyrene, polycarbonate or glass. However, both optical and electrochemical PANI sensors tend to drift (in terms of conductivity or absorbance) in the order of 1% of the background signal per day. This makes daily recalibration necessary prior to measurements. Microtiterplates (micro-well) assays offer an alternative to conventional sensing because calibrators can be placed in one or more wells while actual assays are performed in the residual wells. This is one of the reasons why micro-well assays are commonly employed in routine analysis. In this way, a technique for coating the wells of microtiterplates with polyaniline layers and with polyaniline/enzyme layers is presented by Piletsky et al. [4]. The resulting wells are shown to be useful for assaying enzyme substrates (as exemplified for glucose via pH) and hydrogen peroxide (via the redox properties of the film). Analyte detection is based on monitoring the absorption spectra of the polyaniline, which turn purple as a result of redox processes, or green on formation of acids, by enzymatic reactions. Hydrogen peroxide (a species produced by all oxidases) and glucose (which yields protons on enzymatic oxidation) have been determined in the millimolar to micromolar concentration range.

Recently, Llobera et al. [5] present the characterization and optimization of flexible transducers with demultiplexing properties, suitable for on-chip detection. The micro-system consists of a hollow prism that can be filled with the fluid to be investigated. Two 2D (cylindrical) biconvex lenses modify the optical path before and after propagating through the prism, having parallel beams inside the prism and focusing on the output optical fiber. Light is coupled into the system through a multimode optical fiber inserted into a channel. This channel enables the exact positioning of the fiber with respect to the biconvex lens, conferring the device with a simple, yet effective, self-alignment system. The same applies for the output fiber which enables the collected light to reach the photo-detector. The optimization of the hollow prisms has been done by measuring the absorbance as a function of the concentration for fluorescein and methylorange. Methylorange absorbs blue–green (λ = 490–510 nm) or blue (λ = 435–480 nm) light depending if its molecules are in its acidic or basic form, respectively. When a fixed concentration of methylorange is mixed with buffer with different pH, the fraction of each form of methylorange varies, causing a variation of the absorption spectrum of the solution. The working wavelength used in this work is only absorbed by the basic form (λ = 460 nm), hence, its absorbance will be minimum for low pH values. As the fraction of methylorange in its basic form increases (that is, as the pH increases), the absorbance of the working wavelength will increase. Hence, pH can be measured as a function of the absorbance of the methylorange in its basic form. Results show how the limit of detection (LOD) for fluorescein and methylorange diluted in phosphate buffer can be significantly lowered, by increasing the size of the prism or increasing the total deviation angle (the measurements showed a LOD in the μM range for both species).

Surface plasmon resonance (SPR) is an optical phenomenon caused by charge density oscillation at the interface of two media with dielectric constants of opposite sign, for example a metal and a dielectric. In this way, by tuning the plasmon resonance to a wavelength for which the outer medium is absorptive, a significant variation of the spectral transmittance of the device is produced as a function of the concentration of the analyte. With this mechanism, selectivity can be achieved without the need of any functionalization of the surfaces or the use of recognizing elements, which is a very interesting feature for any kind of chemical sensor or biosensor. Doubly deposited uniform-waist tapered fibers are well suited for the development of these new sensors.

Multiple surface plasmon resonance, obtainable in those structures, can be used for the development of microspectrometers based on this principle [6].

Since the coming of the FTIR spectrometers, infrared spectroscopy has become an indispensable and efficient tool of analytical studies. Nevertheless, classical attenuated total reflection (ATR) or transmission registration need also to collect samples. An alternative method to acquire the infrared spectra consists of using some optical fiber, avoiding then the samplings. For this, the fiber is employed, on the one hand, to transmit the IR beams from the spectrometer to the sample, and on the other hand, as a probe by inserting a part of the fiber, called the sensing zone, into the studied environment. This technique is called fiber evanescent wave spectroscopy (FEWS) for fiber evanescent wave spectroscopy because it is generally considered that the principle of the measurement is based on the presence of evanescent wave around the fiber during the propagation of light into the fiber. A new generation of optical fibers has been developed based on the large transparency domain of an original family of IR chalcogenide glasses transmitting from 2 to about 12 μm.

Fiber-based infrared sensing has been established as an efficient, non-destructive and selective technique for the detection of organic and biological species. This technique combines the benefits of ATR spectroscopy with the flexibility of using a fiber as the transmission line of the optical signal, which allows for remote analysis during field measurements or in clinical environments. The sensing mechanism is based on the absorption of the evanescent electric field, which propagates outside the surface of the fiber and interacts with any absorbing species at the fiber interface. This mechanism is analogous to that observed with an ATR crystal; however the fiber geometry creates a large number of internal reflections which enhances detection sensitivity. The availability of fibers with high infrared transmission in the spectral region between 400 and 4000 cm^{-1} allows one to collect the highly specific vibrational spectrum of organic chemicals and biomolecules. This technique, known as fiber evanescent wave spectroscopy (FEWS), has been applied to the detection of a wide range of chemicals and pollutants.

Since this spectral range comprises the "fingerprint" region of biomolecules, the FEWS technique has also shown promise for biomedical and clinical applications. It can be used as an efficient tool for chemical analysis of bio-fluids [7]. More recently, IR fiber sensors have been applied to monitor the metabolism of live whole cells. This ability to monitor metabolic processes in live cells has interesting potential for the design of bio-optic sensors. Disruption of the cell metabolism can be observed in response to minute amounts of toxicants, which would otherwise be far below the detection limit of IR spectroscopy. In effect, the cells act as a sensitizer for the IR sensor. Additionally, the process of monitoring a cell response permits detection of a wide range of compounds, which may have similar toxicological activities but different molecular structure. Hence biochemicals are detected based on their activity rather than identity, an important distinction critical in the design of sensors which may be challenged with a wide range of analytes to be detected.

Chalcogenide glasses are an ideal choice for the design of IR fiber bio-sensors based on materials properties. Mid-infrared fibers have been used previously for the characterization of live biological samples. An IR fiber optic neurotoxin biosensor was constructed by applying a biologically active cladding to the core of an infrared transmitting chalcogenide fiber [8]. Binding of the surface bound receptor protein was monitored by performing infrared difference spectroscopy on the fiber optic probe before and after its exposure to various concentrations of neurotoxin in solution. Signals measuring conformational changes as a result of these interactions are observed to saturate in agreement with established biochemical kinetics for the receptor. Fiber-optic components are shown to be much more sensitive than bulk optical components in performing these measurements. These fibers have also been functionalized with biological surface coatings such as enzyme films for selective bio-sensing [9]. In order to immobilize glucose oxidase on the surface of such an IR-transparent wave guide, crystalline bacterial cell surface layers (S-layers) were used as a carrier, instead of using silanes as an enzyme coupler as frequently described in the literature. S-layer proteins, which have the capability for self-assembling on suitable surfaces, were cross-linked and further activated with glutaraldehyde before the immobilization procedure. The reactive enzyme layer coating the core of the fiber serves to catalyze chemical reactions specifically when the fiber is used as chemical sensor. The chalcogenide fiber was coupled to a Fourier transform infrared (FT-IR) spectrometer which yielded spectra at various stages of the chemical processes as well as developments of signal bands as a function of time. The fiber was used as ATR element and could provide evanescent-field IR spectra in the range of 4000-800 cm^{-1} of the covering surface film thickness estimated at ~ 40 nm. All experimental surface modifications were carried out in situ in a 12-cm-long flow cell into which the fiber was positioned initially.

Lucas et al. [10] functionalize the surface of chalcogenide Te–As–Se fibers with live human lung cells that can act as sensitizer for the detection on micromolar quantities of toxic agents. First it presents how the hydrophobic behavior of chalcogenide glass affects the spectroscopic properties of chalcogenide fibers. Then it presents initial results on the variation of cell spectra in response to various toxic agents. This study emphasizes the potentials of chalcogenide fibers for the design of cell-based bio-optic sensors.

Near-infrared spectroscopy (NIRS) use depends on the relatively good transparency of biological tissue in the near-infrared range, which allows for transmission of photons through the tissue, so that they can be detected at the exit from the tissue. In particular, oxygenated haemoglobin (HbO2) and deoxy-haemoglobin (Hb) are the dominant absorbing elements between 700 and 1000 nm, and the transmission of light is relatively unaffected by water in the same region. Thus, the near-infrared region of the spectrum is the most favourable to the optical measurement of these parameters, so that NIRS provides a non-invasive, non-ionizing means to monitor total haemoglobin concentration (HbO2 + Hb) that is considered as total blood volume (HbT or V) as well as oxygen saturation in the living tissue. Optic fiber probes were used as the optical head of a novel, highly sensitive near-infrared continuous wave spectroscopy (CW-NIR) instrument. This prototype was designed for non-invasive analysis of the two main forms of haemoglobin [11].

3. REFLECTANCE MEASUREMENTS

Interest has been shown in the detection of free radicals in view of the evidence implicating a primary or secondary role for them in the initiation or progression of many diseases, the majority of which are characterized by an inflammatory reaction. Naughton et al. [12] describe the development of a fibre optic sensor, based on immobilized nitrophenol that is of potential use for the continuous monitoring of OH radical production. This reflectance based sensor incorporates an OH radical sensitive chromophore, affording a decrease in its reflectance spectrum upon attack by this extremely reactive oxygen-derived radical attributable to the formation of nitrocatechol. Nitrophenol was immobilized onto XAD-7 methacrylate beads. Subsequently the beads were attached to the distal end of a polymethylmethacrylate fiber optic. Nitrocatechol, generated from the attack of OH radical on nitrophenol, exhibits a strong absorption band in the visible region of the electromagnetic spectrum (λ (max) = 510 nm).

Dyr et al. [13,14] evaluated the feasibility to follow the enzymatic conversion of surface-bound fibrinogen by thrombin to fibrin monomer and possible complex formation between surface-bound fibrin monomer and fibrinogen in solution using a SPR sensor. For the investigations of optical properties of immobilised molecular layers, the Kretschmann geometry of the attenuated total reflection (ATR) method was utilised. Density of fibrinogen bound to the surface depended on the concentration of fibrinogen in solution during the adsorption process. A fibrinogen monolayer was always formed. Fibrinogen in solution did not bind to surface-bound fibrinogen. Bound fibrinogen converted by thrombin to fibrin monomer interacted (rather slowly) with fibrinogen in solution. The rate of adsorption depended upon immobilised fibrin monomer density, fibrinogen concentration in solution, and on the presence of calcium ions. At low fibrin monomer density, the second layer was formed that contained about the same amount of protein as the first layer, at higher fibrin monomer concentration less than one molecule of fibrinogen per molecule of fibrin monomer was captured.

A novel approach for the detection of molecular interactions in which a colorimetric resonant diffractive grating surface is used as a surface binding platform is proposed by Cunningham et al. [15]. The grating, when illuminated with white light, is designed to reflect only a single wavelength. When molecules are attached to the surface, the reflected wavelength is shifted due to the change of the optical path of light that is coupled into the grating. By linking receptor molecules to the grating surface, complementary binding molecules can be detected without the use of any kind of fluorescent probe or particle label. The detection technique is capable of resolving changes of ~ 0.1 nm thickness of protein binding, and can be performed with the grating surface either immersed in fluid or dried. The readout system consists of a white light lamp that illuminates a small spot of the grating at nominally normal incidence through a fiber optic probe, and a spectrometer that collects the reflected light through a second fiber, also at normal incidence. Because no physical contact occurs between the excitation/readout system and the grating surface, no special coupling prisms are required and the grating can be easily adapted to any commonly used assay platform, such as microtiter plates and microarray slides. A single spectrometer reading may be performed in several milliseconds, thus it is possible to quickly measure a large number of molecular interactions taking place in parallel upon a grating surface, and to monitor reaction kinetics in real time.

Based on the principle of multiple-reflection in white-light interferometry, an optical fiber immobilized with protein molecule (antigen or antibody) is adopted as a high sensitivity biosensor to accomplish the real time monitor of the immunoassay [16-18]. The experimental results of rabbit-IgG with anti-rabbit-IgG immunoreaction show that by the aid of this biosensor, a biolayer change of approximately 0.70 nm in optical thickness can be detected by measuring the reflected spectrum phase shifting between the two interfaces of biolayer. Compared with others, this method has advantages, such as simple structure, lower cost, high sensitivity and reliability, label-free in immunoassay and direct monitor the immunoassay. This biosensor is easy to be integrated as a BIAcore (biomolecular interaction analysis core) for parallel direct detection of antigen-antibody interactions.

SPR biosensor has been widely used in the last 10 years to analyze the affinity between ligand and analyte [19]. The SPR condition is usually achieved in the attenuated total reflection (ATR) geometry by using prism-coupling optics. This optical setup is usually bulky and is incompatible with micrometer-sized probes. In place of the conventional SPR, interest in localized plasmon resonance (LPR) has grown in recent years, since LPR is quite compatible with such microsensors. The LPR of gold appears as an absorption band at about 550 nm in the presence of thin molecular layers on gold. In this condition, the LPR produces a large enhancement of the electric field around small gold particles or roughness of a gold surface, so that it is applicable in a highly sensitive affinity sensor of size less than microns. Kajikawa et al. propose a simple and sensitive optical sensing method for biological applications [20,21]. Since gold behaves as a dielectric with a large extinction index under blue or violet light, presence of a transparent surface layer on gold produces a large decrease in the reflectivity of the gold surface due to multiple reflections in the surface layer. Their call this phenomenon anomalous reflection (AR) of the gold surface. AR is applicable to affinity biosensors based on fiber optics, so that inexpensive and disposable micrometer-sized biosensor probes are possible based on this technique. Here, the authors demonstrate the application of AR to real-time measurements of the adsorption process of octadecanethiol on gold and the affinity of streptavidin to a biotin-labeled monomolecular layer on gold.

Misiakos et al. [22] describe an optical affinity sensor based on a monolithic optoelectronic transducer, which integrates on a silicon die thin optical fibers (silicon nitride) along with self-aligned light-emitting diodes (LED) and photodetectors (silicon p/n junction). The LEDs are optically coupled to the corresponding photodetectors through silicon nitride fibers. Specially designed spacers provide for the smooth bending of the fiber at its end points and toward the light source and the detector ensuring high coupling efficiency. The transducer surface is hydrophilized by oxygen plasma treatment, silanized with (3-aminopropyl)triethoxysilane and bioactivated through adsorption of the biomolecular probes. The use of a microfluidic module allows real-time monitoring of the binding reaction of the gold nanoparticle-labeled analytes with the immobilized probes. Their binding within the evanescent field at the surface of the optical fiber causes attenuated total reflection of the waveguided modes and reduction of the detector photocurrent. The biotin-streptavidin model assay was used for the evaluation of the analytical potentials of the device developed. Detection limits of 3.8 and 13 pM in terms of gold nanoparticle-labeled streptavidin were achieved for continuous and stopped-flow assay modes, respectively. The detection sensitivity was improved by silver plating of the immobilized gold nanoparticles, and a detection limit of 20 fM was obtained after 20-min of silver plating. In addition, two different analytes, streptavidin and antimouse IgG, were simultaneously assayed on the same chip demonstrating the multianalyte potential of the sensor developed.

Recently, an optical fiber biosensor is introduced herein, which could directly detect biological interaction such as immunoreactions of antigens and antibodies without destroy the biolayer [23]. The test is based on the theory of multilayer-reflection principle in white-light interferometry. When immunoreactions occur, the reflected spectrum phase shifts. Immunoreactions could be detected by means of reflected spectrum phase shifting, or by biolayer thickness changing. Continuously detecting of thickness changing on a fractional nanometer scale with subsecond repetition times is allowed in this system. The detecting system has high sensitivity, high precision, and high speed, cost effective and working on a high reliability. The bioprobe is easy integrated as a BIAcore. The system and the experimental results on the reaction of rabbit-IgG with anti-rabbit-IgG are described in this work.

4. FLUORESCENCE MEASUREMENTS

Fluorescence techniques provide sensitive detection of biomolecules. Furthermore, since fluorescence intensity is proportional to the excitation intensity, even weak signals can be observed. In last decade reagentless fiber-based biosensors have been developed. These biosensors are capable of detecting changes in cell behaviour, metabolism and cell death when exposed to toxic agents.

Fluorescence measurements are not used as often as absorbance and reflectance with enzyme optical fiber-based biosensors, as it is not common for enzyme reactions to produce fluorescent products or intermediates. Most fluorescence techniques employ a fluorescent dye to indirectly monitor formation or consumption of a transducer.

Several enzyme-catalysed reactions involve the production or consumption of fluorescent species. NAD^+ (nicotinamide adenine dinucleotide)-dependent reactions, catalysed by a group of dehydrogenase enzymes, are some of the notable examples of this chemical phenomenon. In the general dehydrogenase reaction:

$$\text{Substrate} + \text{NAD(P)}^+ \xrightarrow{\text{dehydrogenase}} \text{product} + \text{NAD(P)H}$$

NADH or NADPH is produced, which can be detected using fibre optics through its fluorescence at λ_{ex} = 350 nm, λ_{em} = 450 nm. Several optical sensors have thus been produced by integrating different dehydrogenase or oxidoreductase enzymes with optical fibres. These sensors are listed in Table 1.

A fiber optic-based biosensor which integrates a novel array of optical and electrical components, together with long fused silica fibers and proteins for detection of analyte in solution has been proposed by Anderson et al. [98]. The optical fiber core near the distal end is tapered and coated with either antibodies or DNA binding proteins. Assays are performed by flow of a solution containing the fluorescently tagged ligand molecules over the coated fiber. Within seconds, analyte recognition occurs and a fluorescence signal is transmitted back up the fiber. Applications for the biosensor include clinical diagnostics, pollution control, and environmental monitoring.

The use of DNA as a selective recognition element in biosensor design is a new and exciting area in analytical chemistry. Experiments have been completed wherein ssDNA was covalently immobilized onto quartz optical fibers to successfully demonstrate the basis for development of an optical biosensor for DNA. The non-optimized configuration described was able to detect femtomolar concentrations of cDNA with an analysis time of less than 1 h. These experiments indicate that biosensors which can selectively detect genetic material from biological may now be created, with advantages of low detection limits, reasonable analysis times, highly stable biorecognition elements, and regenerability. An alternative approach which may be used to create biosensors with long term chemical stability takes advantage of the stability of DNA. With the recent advent of DNA probe technology, a number of selective oligomers which interact with the DNA of important biological species, for instance salmonella, have been identified. These have been used to provide a new type of selective biorecognition element which is highly selective, stable, and can be easily synthesized in the laboratory as compared to other chemically synthesized biorecognition elements. As a result, species specific DNA probes may be exploited for biosensor development.

Piunno et al. [99,100] reports one of the first biosensors for direct analysis of DNA hybridization by use of an optical fiber. ssDNA was covalently immobilized onto optical fibers by first activating the surface of the quartz optical fiber with a long chain aliphatic spacer arm terminated in a 5′-0-dimethoxytrityl-2′-deoxyribonucleoside, followed by automated solid-phase DNA synthesis. Detection of dsDNA at the fiber surface after hybridization between immobilized ssDNA and cDNA was achieved by exposing the complex to an ethidium bromide solution followed by washings with hybridization buffer solution. The ethidium cation (3,8-diamino-6-phenyl-5-ethyl-phenanthridium) is a fluorescent compound which strongly associates with dsDNA by intercalation into the base stacking region and, in some cases, the major groove of the double helical structure. It is shown that the fluorescence response of the ethidium cation can be monitored in a total internal reflection configuration along an optical fiber to quantify the presence of dsDNA at the surface of the fiber, with the fluorescence intensity being directly proportional to the amount of cDNA initially present in solution. This approach may be refined and used for rapid identification and quantitation of the presence of microorganisms such as pathogenic bacteria and viruses in bodily fluids, food and feed commodities, and may also find application in screening for genetic disorders.

Other fiber optic DNA biosensors and also other fiber optic biosensors based on fluorescence measurements are summarised in Table 1.

Optical biosensors based on fluorescence detection often use the combination of a fluorescent bioreceptor associated with an optical transducer. Fluorescent biosensors may also be obtained by immobilizing whole cells on the surface of a sensor layer. This bioactive layer is usually placed in front of the tip of an optical fibers bundle to generate a fluorescence signal. The optical fibers are required to send the excitation radiation to the fluorescent bioelement and convey the fluorescence radiation up to a fluorimeter. When algal cells are deposited on the surface of an opaque support, they are placed at some distance to the optical fibers bundle to allow the fluorescence radiation emitted by the illuminated area to be collected properly. In order to improve the simplicity and reliability of fluorescence based biosensors, optically translucent supports are used because their optical properties enable detection of fluorescence emitted by the algal cells even if they are entrapped in the bulk of the translucent support. Silica matrixes have interesting properties including optical translucence, biocompatibility and chemical inertness. The design flexibility of sol–gel technique and ease of fabrication can fulfil to create the support with structural and chemical features that could be compatible with biomaterials. While immobilization of enzymes and whole cells in sol–gel is a well-known technique in biosensor application, it is not yet used for the construction of vegetal-cell optical biosensors. On the other hand, when an enzyme is immobilized, a substrate is required and a fluorescent indicator may be added to enable optical detection of the analyte. In the case of detection based on chlorophyll fluorescence, those reagents are not necessary, and a reagentless biosensor can be constructed for toxic chemicals determination. Recently, Nguyen-Ngoc and Tran-Minh investigated translucent matrixes obtained from a sol–gel process for entrapment of fluorescent biomaterials in order to improve the fluorescent biosensor fabrication. Vegetal cells as bioreceptors have been tested for herbicides determination [101].

Also, a single, compact probe containing a manifold of chemical and biological recognition principles that could all be simultaneously and independently interrogated forms an excellent foundation for the development of clinical and environmental total analysis systems. Imaging fibers consisting of a fused fiber bundle that can each be optically addressed independently are very useful for optical imaging applications. In this sense, optical imaging fibers with micrometer-sized wells were used as a sensing platform for the development of microarray optical ion sensors based on selective bulk extraction principles established earlier for optodes. Uniform 10 mm sized microspheres based on plasticized poly(vinyl chloride) containing various combinations of ionophores, fluoroionophores and lipophilic ion-exchangers were prepared for the detection of sodium, potassium, calcium and chloride, and deposited onto the wells of etched fiber bundles. The fluorescence emission characteristics of individual microspheres were observed from the backside of the fibers and were found to selectively and rapidly change as a function of the sample composition [102].

Table 1. Fiber optic biosensors based on fluorescence measurements.

Applications	Remarks	Ref.
Lactate, pyruvate, ethanol	Using confined macromolecular nicotinamide adenine dinucleotide derivatives; flow-injection analysis (FIA)	24
Lactate, phenylpyruvate, pyruvate, formate, ethanol and mannitol concentrations in the range of 50-1500 μmol l^{-1}	Miniaturized fluorimeter with a low-volume membrane reactor and utilizes FIA; two types of membrane reactor: a) enzymes and a macromolecular nicotinamide adenine dinucleotide coenzyme derivative of polyethyleneglycol are retained in the direct vicinity of the sensor tip; b) solid-supported liquid membrane used to retain native coenzyme, while substrates and products transported via selective carriers through an organic membrane barrier	25
Lactate	Based on bacterial cytoplasmic membranes (CPM) as the biological recognition element and an oxygen sensitive dye layer as the transducer; CPMs from bacteria with an induced lactate oxidase system are adsorbed onto a cellulose disk that is fixed mechanically over an oxygen sensitive siloxane layer on the distal end of an optical fiber	26
Ethanol, mannitol	FIA; alcohol dehydrogenase + lactate dehydrogenase for ethanol ; mannitol dehydrogenase + lactate dehydrogenase for mannitol	27
Ethanol, urea	For ethanol increase of NADH fluorescence during the enzymatic conversion of ethanol using an enzyme system of ADH and AlDH; for urea based on a pH sensor measuring the change in fluorescence of the pH indicator fluorescein; both sensors are suitable for controlling fermentation processes when ethanol and urea are produced or consumed; software package FIACRE developed to run three different FIA processes simultaneously; it is applied to a model fermentation with Enterobacteriaceae Proteus vulgaris controlled by an urea optode	28
Glutamate	Enzyme layer composed of glutamate dehydrogenase (GDH) and glutamatepyruvate transaminase (GPT) for produce NADH at the tip of a fiber-optic probe; GDH catalyzes the formation of NADH, and GPT drives the GDH reaction by removing a reaction product and regenerating glutamate; optimal response obtained a pH 7.4 Tris-HCl buffer at 25 °C in 4 mM NAD^+ and 10 mM L-alanine	29
Glutamate	GDH directly immobilized onto optical fiber probe surface through covalent binding mechanisms; optical fiber surface initially activated by silanization, which adds amine groups to the surface; then affix functional groups CHO to the optical fiber surface by employing a bifunctional cross-linking agent, glutaraldehyde	30
Glutamate	Comprised of an L-glutamate-sensitive fluorescent gel, spin-coated onto the tip of an optical imaging fiber; gel composed of L-glutamate oxidase (GLOD); a pH-sensitive fluorescent dye, SNAFL; and poly(acrylamide-co-N-acryloxysuccinimide) (PAN); NH_3 is liberated from the interaction	31

	of L-glutamate with GLOD, which reversibly reduces the emitted fluorescence signal from SNAFL	
Fructose, gluconolactone, glucose	Glucose-fructose oxidoreductase; permeabilized cells confined in front of the fiber tip and changes in the NAD(P)H pool of membrane-bound enzyme complexes inside the cells monitored during enzymatic reactions	32
Fructose, glucose	FIA; using glucose-fructose-oxidoreductase isolated from Zymomonas mobilis confined in a measurement cell behind an urtrafiltration membrane	33
Acetylcholinesterases (AChEs)	By immobilizing fluorescein isothiocyanate (FITC)-tagged eel electric organ AChE on quartz fibers and monitoring enzyme activity; pH-dependent fluorescent signal generated by FITC-AChE, present in the evanescent zone on the fiber surface, was quenched by the protons produced during acetylcholine (ACh) hydrolysis	34
Nicotinic acetylcholine receptor	Using three fluorescein isothiocyanate (FITC)-tagged neurotoxic peptides that vary in the reversibility of their receptor inhibition: α-bungarotoxin, α-Naja toxin, and α-conotoxin	35 36
Ethanol vapor	By immobilizing alcohol oxidase (AOD) onto a tip of a fiber optic oxygen sensor with a tube-ring, using an oxygen sensitive ruthenium organic complex (λ_{exc}=470 nm; λ_{em}=600 nm); reaction unit for circulating buffer solution applied to the tip of the device; after experiment in liquid phase, the sniffer-device was applied for gas analysis using a gas flow measurement system with a gas generator; optical device applied to detect the oxygen consumption induced by AOD enzymatic reaction with alcohol application	37
Glucose	Biochemical assay based on homogeneous singlet/singlet energy transfer affinity assay; sensor probe indirectly measures glucose concentrations from the level of fluorescence quenching caused by the homogeneous competition assay between TRITC labeled concanavalin A (receptor) and FITC labeled Dextran (ligand)	38
Glucose	Based on coupling the swelling of a polymer gel to a change in fluorescence intensity; fluorophore, an amine functional group and enzyme glucose oxidase were each incorporated into a crosslinked polymer gel, which was formed on the end of a fibre optic rod; while the amount of fluorophore remains constant, the gel volume changes in response to a change in the ionization state of the amine moiety; change related to glucose concentration	39
Anticonvulsant drug phenytoin (5,5-diphenylhydantoin) in human blood and plasma	Using a self-contained fiber-optic immunosensor; signal depended on the degree of energy transfer from B-phycoerythrin labeled with phenytoin to Texas Red labeled anti-phenytoin antibody; Dextran 70K added to system to equalize oncotic pressure across the encapsulation membrane	40
Simultaneous analysis of multiple DNA sequences	A bundle of optical fibers assembled with each fiber carrying a different oligonucleotide probe immobilized on its distal end; hybridization of	41

	fluorescently labeled complementary oligonucleotides to the array monitored by observing increase in fluorescence that accompanied binding; approach enables fast (<10 min) and sensitive (10 nM) detection of multiple DNA sequences simultaneously	
DNA sensor	Using Lagmuir-Blodgett technique cinnamoylbutylether-cellulose monolayers transferred onto optical fibers or planar waveguides; films served as matrices for immobilization of biotinylated oligonucleotides via streptavidin; specificity of streptavidin layer or following bounded nucleic acid molecules controlled by enzyme- linked immunosorbent assay	42
16-mer oligonucleotides in DNA hybridization assays	Based on fluorescence excitation and detection in the evanescent field of a quartz fiber; biotinylated capture probe immobilized on the fiber surface via avidin or streptavidin	43
DNA sensor	Two methods for covalent immobilization of single-stranded DNA onto optical fibers: a) fused silica optical fibers derivatized with γ-amino-propyltriethoxysilane followed by extension with 12-nitrododecanoic acid via amide formation; terminal nitro moiety then reduced to amine to provide a hydrophobic spacer molecule suitable for covalent attachment of an oligonucleotide; b) fibers activated with 3-glycidoxypropyltrimethoxysilane and a hydrophilic spacer arm composed of 1-17-diamino-3,6,9,12,15-pentaoxaheptadecane covalently attached; oligonucleotides (dT20) then bound to either type of spacer arm through a phosphoramidate bond between amine terminus of the spacer molecule and 5'-phosphate of the DNA via a carbodiimide mediated condensation reaction	44
Specific determination of femtomolar DNA oligomers	13mer oligonucleotides attached to core of a multimode fibre; complementary sequence detected by use of a fluorescent double strand specific DNA ligand (YOYO and PicoGreen); evanescent field employed to distinguish between bound and not bound species; template DNA-oligomer immobilized either by direct coupling to activated sensor surface or using avidin-biotin bridge	45
Detection of genomic target sequences from Escherichia coli	A small portion of the LacZ DNA sequence is the basis for selection of DNA probe molecules that are produced by automated nucleic acid synthesis on the surface of optical fibres; fluorescent intercalating agents are used to report the presence of hybridization events with target strands	46
To fabricate a mercury biosensor	Using genetically modified Escherichia coli strain, containing lacZ reporter gene fused to heavy metal-responsive gene promoter zntA; plasmid carrying gene coding for enhanced cyan fluorescent protein also introduced into sensing strain to identify cell locations in array; single cell lacZ expression measured when array was exposed to mercury and a response to 100nM Hg^{2+} could be detected after 1-h incubation time; optical imaging fiber-based single bacterial cell array is flexible and	47

	sensitive biosensor platform that can be used to monitor the expression of different reporter genes and accommodate a variety of sensing strains	
Detection of hybridization of nucleic acids	Probe DNA was dT20, and target was Fluorescein-labeled non-complementary (dT20) or complementary (dA20) oligonucleotide; chronofluorimetric monitoring of adsorption and hybridization processes used to investigate oligonucleotide films of different density, in different salt concentrations, at temperatures of 25 and 40 °C, with concentration of the target DNA being 0.005-0.1 μM.	48
Thrombin and also a similar strategy could be used to study other analytes such as protein and small molecules	By aptamer connected with molecular beacon that consist of an oligonucleotide sequence containing complementary sequence sections at either end; these two sequence containing segments base pair with each other to form a hairpin shaped loop structure, the fluorophore and quencher were attached at 5'- and 3'-end of molecular beacon, respectively	49
Hybridization assays	Experiments were done to examine the effects of immobilization density and solution conditions on sensitivity, selectivity and dynamic range of hybridization assays done using a fiber optic nucleic acid biosensor based on TIRF	50
Hybridization assays	On the evanescent field excitation of fluorescence from surface-bound fluorophores; evanescent field made accessible through use of long, adiabatically tapered single-mode fiber probe; laser diode with a 785-nm wavelength used in a pulsed mode of operation to excite fluorescence in the tapered region of a fiber probe using the near-infrared fluorophore IRD 41	51 52
Determination of nucleic acid hybridization	1-hydroxyethyl-4-[(3-methyl-6-methoxybenzothiazole-2-ylidine) methine] quinolium bromide, a derivative of thiazole orange, use as a fluorescent indicator of DNA hybrid formation on a fiber optic genosensor platform	53
By using dinitrophenyl (DNP) hapten-loaded liposomes, concentrations as low as 500-fold diluted anti-DNP antibody and 0.76 CH_{50} ml^{-1} of complement were detected	Agar gel-immobilized liposomes containing carboxyfluorescein attached to the tip of optical fiber; complement-mediated immunolysis of liposome was fluorometrically detected through the fibre	54
Competitive binding immunofluorescence assay	Protein A covalently immobilized on activated sensing tip surface of quartz fiber or on activated surface of glass cover slips; surface then incubated with rabbit IgG to reach a reasonable binding density and homogeneous distribution for adequate detectability; sensor exposed to FITC labeled and unlabeled goat anti-rabbit IgG; fluorescence of sensor bound analyte, excited by Argon ion laser 476 nm line and emitted from sensing tip or sensing spot on the sample slips, is coupled back into the fiber	55
Quantifying serum antiplague antibody	Fiber probes with immobilized antiplague IgG coated with F1 antigen and then incubated with serum samples, quantity of antiplague serum	56

	antibodies which bound to the probe surface was then determined with fluorescent rabbit anti-human IgG	
Protein C in a plasma sample; immunosensor was also tested with human serum albumin and human plasma and with and without convective flow	Optimal primary antibody concentration is 65 μg ml^{-1}; its leaching as minor, stabilized within 3 days, and the sensor sensitivity change after 30 days of storage was minor; sample and secondary antibody incubation times reduced from 10 to 5 and from 5 to 3 min, respectively, while maintaining a reasonable signal-to-noise level	57
Antibody immobilization	Several heterobifunctional crosslinkers compared to N-succinimidyl 4-maleimidobutyrate for their ability to immobilize active antibodies onto glass cover slips at a high density	58
Proteins	Matrix uses a gold colloid monolayer attached to an end of a fiber as a substrate for protein attachment	59
Zinc	Sensor transduces the specific recognition of the ion by an enzyme (carbonic anhydrase) as a change in the fluorescence of an inhibitor which binds to the zinc in the active site	60 61
Zinc	Synthesis of difluorofluorescein monocarboxaldehyde platform and use for preparing ZP8, a new member of the Zinpyr family of neuronal Zn^{2+} sensors	62
Rapid and sensitive fluoroimmunoassays for biologic analytes	Using a laboratory breadboard device that employed a large 514 nm argon ion laser; also portable, multichannel biosensor was developed which uses 635 nm laser diodes, the assays were converted to use the cyanine dye, Cy5; detection antibodies were labeled with Cy5 and assays performed to detect F1 antigen of Yersinia pestis and the protective antigen of Bacillus anthracis	63
Applications of sol-gel encapsulated phycobiliproteins	Light transducing phycobiliproteins encapsulated in optically transparent sol-gel matrices	64
Proteins	A model biorecognition element-reporter group couple consisting of human serum albumin that was site-selectively labelled at Cys 34 with iodoacetoxy-nitrobenzoxadiazole; labelled protein encapsulated into sol-gel derived materials	65
Acetylcholine	Sensor consists of a pH-sensitive fluorescent indicator encapsulated with enzyme in a sol-gel network on a glass cap that can be fixed on an optical fiber and then integrated with a flow-through reactor for continuous monitoring	66
Protein A	Effectiveness of protein A, an immunoglobulin binding protein, for antibody immobilization on the surface of fiber probes	67
Organophosphorus pesticide	Sol-gel crystals derived from tetramethyl orthosilicate doped with cholinesterase using microencapsulation	68
Fiber optic cocaine biosensor	Replacement of quartz halogen lamp of fluorometer with high brightness blue-light-emitting diode enhanced evanescent excitation of	69

	benzoylecgonine-fluorescein bound to mAbs immobilized on quartz fibers and gave stronger optical signals	
Penicillin or glucose in the presence of a concurrent pH or O_2 change, respectively	Biosensors are fabricated by site-selective photodeposition of analyte-sensitive polymer matrices on optical imaging fibres	70
Oxygen sensing in non-aqueous media	Using porous glass with covalently bound luminescent Ru(II) complexes in methanol, chloroform, toluene, cyclohexane, and n-hexane	71
Cellular detection of nitric oxide	Sensors incorporate cytochrome c', a hemoprotein known to bind nitric oxide selectively; cytochrome c' is labeled with a fluorescent reporter dye	72
New detection instrument for chemical/biological fluorescence lifetime-based sensors	Dual closed-loop optoelectronic auto-oscillatory detection circuit: primary loop consists of a fluorescence excitation light source, a fiber-optic delay line (with a gap for placement of a fluorescent sensor), an electronic phase shifter, a photo-detector, and a resonance-type RF amplifier; secondary loop consists of a long-wavelength-pass optical filter, multimode fiber, a PMT, and an electronic phase detector (which is connected to the phase shifter of the primary loop)	73
Picric acid	By replacing the hydrogen of the 4-amino group of a 4-amino-1,8-naphthalimide derivative with an N-acryloxyethyl group, the fluorophore has been covalently immobilized on an optical sensor surface by UV photopolymerization	74
Development of a multi-pathogen, immunoassay-based, fiber optic detector	For attaching quantum dots (QDs) to antibodies for use in biodetection applications: synthesized CdSe/ZnS core-shell QDs of differing size, functionalized their surfaces with several types of organic groups for water solubility, and covalently attached these functionalized QDs to rabbit anti-ovalbumin antibody protein	75 76
Application of evanescent wave sensing to a high-sensitivity fluoroimmunoassay of the clinically important human enzyme creatine kinase (CK) isoenzyme MB form (CK-MB)	Using a green helium-neon laser based fluorimeter and the antibody conjugated fluorophor B-phycoerthrin	77
Fluorescein	Comparison between evanescent wave generation and distal-face generation of fluorescent light	78
Evanescent wave fiber optic biosensor	Combination tapered fibers designed for improved signal acquisition	79
Antibody/antigen binding	Ray-tracing determination of evanescent wave penetration depth in tapered fiber optic probes	80
Evanescent wave biosensor	Fluorescent signal acquisition from step-etched fiber optic probes or from tapered fiber optic probes	81 82 83
Evanescent wave immunoprobe with high bivalent antibody activity	Using both kinetic methods and optical determinations of bound antigen, there are 2.4×10^{11} active antibodies per cm^2 probe area; 75% of the active	84

	antibodies are in bivalent form, with both binding sites capable of binding antigen	
A pH sensor array and an acetylcholine biosensor array	Combined imaging and chemical sensing using a single optical imaging fiber	85
Novel single mode tapered optical fibre loop biochemical sensor	Analyte labelled secondarily with a fluorescent dye, fluorescein isothiocyanate or tetramethyl rhodamine; when excited by input argon ion laser light from the near end of the taper, generated fluorescence is coupled into the guided mode of the fibre and collected at the far end of the taper	86
Portable multichannel fiber optic biosensor for field detection	Using a novel optical fiber bundle jumper for exciting and collecting fluorescence emission from the evanescent wave fiber optic probes; single fiber in the center of bundle couples laser excitation into the sensor probe, while the surrounding fibers collect the returning fluorescent emission light	87
Tapered single-mode optical fiber evanescent coupling	Investigation of two different geometries of nonadiabatic, tapered single-mode optical fiber	88
Design and application of fiber-optic evanescent wave biosensor	Using a red laser diode at 636.85 nm for exciting Cy5 fluorescent dye	89
Fluorescent coagulation assay for thrombin	Using a fibre optic evanescent wave sensor; coagulation of solution phase fluorescently labelled fibrinogen to unlabelled fibrinogen bound to surface of fibre optic observed in real time by evanescent wave sensor	90
Multi-probe fiber optic evanescent wave biosensor and its characterization	Using laser diode at 635 nm wavelength as light source; biosensor can interrogate five individual fibers at one time; assays for five separate analyses can be conducted on the same sample	91
Sencil™ project: Development of a percutaneous optical biosensor	Designed to be minimally invasive, disposable and easily readable to make frequent measurements of various analytes in vivo over a period of 1-3 months; uses photonic sensing of a chemical reaction that occurs in a polymer matrix bound to the internal end of a chronically implanted percutaneous optical fiber	92
Effect of taper geometries and launch angle on evanescent wave penetration depth in optical fibers	Evanescent wave penetration depths of the order of the size of living cells have been achieved by optimizing the parameters relating geometry of tapered fibers; theoretical results for variation of penetration depths of evanescent waves as a function of taper ratios of tapered fiber in various taper geometries, taper length and launch angles	93
Structure study of fiber optical evanescent wave biosensor	In order to improve its detection limit and field applicability, the evanescent wave strength of fiber probe and the compactness of the system are studied	94
Evanescent wave optic fiber biosensor	FITC-labeled goat anti-human IgG were covalently immobilized on 3-aminopropyltriethoxysilane and glutaraldehyde to modify the distal end of a fiber-optic probe	95
For detecting biomarkers in biosamples	A nanogold particle (NGP), when placed at an appropriate distance from a	96
	fluorophore, can effectively enhance the fluorescence by transferring the free electrons of the fluorophore, normally used for self-quenching, to the strong surface plasmon polarization field of the NGP	
A novel type of fiber optic biosensor	First microstructured polymer optical fiber (mPOF) fabricated from Topas cyclic olefin copolymer	97

5. CHEMILUMINESCENCE MEASUREMENTS

A molecular assembly in which a conjugated polymer is interfaced with a photodynamic protein is described by Ayyagari et al. [103]. The conjugated polymer, functionalized with biotin, is designed such that it can be physisorbed on or chemically grown off a glass surface. The streptavidin-derivatized protein is immobilized on the biotinylated polymer matrix through the strong biotin-streptavidin interactions. The assembly, built on the surface of an optical fiber or on the inside walls of a glass capillary form an integral part of a biosensor for the detection of environmental pollutants such as organophosphorus-based insecticides. The protein in the system can be replaced by any biological macromolecule of interest. These authors study one specific case, the inhibition of the enzyme alkaline phosphatase. The enzyme catalyzes a reaction producing an intermediate compound that chemiluminesces, and the chemiluminescence signal is monitored to detect and quantify insecticides such as paraoxon and methyl parathion. Preliminary results indicate ppb level detection with response time less than 1 minute.

Chen et al. [104-106] report a technique to immobilize a multilayer enzyme assembly on an optical fiber surface. A multilayer of an enzyme, alkaline phosphatase, was immobilized by chemical cross-linking on an optical fiber surface. Chemiluminescence, ellipsometry, and surface plasmon resonance were used to characterize the structure and activity of the assembly. A chemiluminescence-based fiber optic biosensor utilizing this immobilization technique has been developed for the detection of organophosphorus-based pesticides.

A fiber optic biosensor based on the electrochemiluminescence of luminol has been developed for glucose and lactate flow injection analysis by Marquette and Blum [107,108]. The electrochemiluminescence of luminol was generated using a glassy carbon electrode polarised at +425 mV vs. a platinum pseudo-reference electrode. After optimisation of the reaction conditions and physicochemical parameters influencing the sensor response, the measurement of hydrogen peroxide could be performed in the range 1.5

pmol-30 nmol. Glucose oxidase or lactate oxidase, were immobilised on polyamide and collagen membranes. With collagen as the enzymatic support, the detection limits for glucose and lactate were 60 pmol and 30 pmol, respectively, whereas with the enzymatic polyamide membranes, the corresponding values were 150 pmol and 60 pmol.

A chemiluminescence biosensing system for antioxidants was developed by Palaoran et al. [109] based on luminol and hematin co-immobilized on a cellulose membrane disc. The concentration of the antioxidant was quantified through the measurement of the inhibition of the chemiluminescence emitted when hydrogen peroxide was introduced into the reagent phase. The instrumentation employed in the measurement was a fabricated luminometer employing optical fibers and a UV- enhanced photodiode transducer. The minimum detectable concentration was 100 μM, and the response time was less than 60 seconds.

The enhanced chemiluminescence (CL) reaction of the luminol-H_2O_2-horseradish peroxidase (HRP) system with immobilized HRP using microencapsulation in a sol-gel matrix has been used to develop a biosensor for p-iodophenol, p-coumaric acid, 2-naphthol and hydrogen peroxide [110]. The detection limits obtained for p-iodophenol, p-coumaric acid and 2-naphthol were 0.83 μM, 15 and 48 nM, respectively. Direct enzyme immobilization onto the end of the optical fibre permits the construction of a remote enhanced CL biosensor. This remote biosensor has been applied to hydrogen peroxide assay (detection limit 52.2 μM).

In other study, an optical fiber bienzyme sensor based on the luminol chemiluminescent reaction was developed and demonstrated to be sensitive to glucose [111]. Glucose oxidase and horseradish peroxidase were co-immobilized by microencapsulation in a sol-gel film derived from tetraethyl orthosilicate. The calibration plots for glucose were established by the optical fiber glucose sensor fabricated by attaching the bienzyme silica gel onto the glass window of the fiber bundle. The linear range was 0.2-2 mM and the detection limit was approximately 0.12 mM.

Recently, Magrisso et al. [112] describe the construction of a novel computerized multi-sample temperature-controlled luminometer for a fiber array-based biosensor to monitor circulating phagocyte activity. It can perform simultaneously integral measurements of chemiluminescence emitted from up to six samples containing less that 0.5 μl whole blood while the samples and detector do not change their position during the measurement cycle. The optical fibers in this luminometer are used as both light guides and solid phase sample holders. The latter feature of the instrument design simplifies the assessment process of both the extra-cellular and the intra-cellular parts of the phagocyte-emitted chemiluminescence using the same system. This new technology may find use in a wide range of analytical luminescence applications in biology, biophysics, biochemistry, toxicology and clinical medicine.

6. BIOLUMINESCENCE MEASUREMENTS

A biosensor associates a bioactive sensing layer with a suitable transducer giving a usable output signal. Although the selective molecular recognition of the target analyte can theoretically be achieved with various kinds of affinity systems, in most biosensors, enzymes are concerned. The biocatalysts are generally immobilized on an artificial support placed in close contact with the transducer. Covalent immobilization is preferable because it prevents the leakage of enzymes from the support. This can now be easily achieved with the availability of commercially preactivated membranes. Sensing layers prepared with such membranes were used for the development of amperometric as well as luminescence-based fiber-optic biosensors. In addition to the target analyte, enzymatic reactions generally involve one or two other substrates, which must be added to the reaction medium when an enzyme based biosensor is operated. The sensing scheme, and therefore the transducer associated with the sensing layer, determines the sensitivity of an enzyme-based sensor. Due to the peculiar nature of one of the reaction products, i.e., light, which can be detected at a very low level, bioluminescence and chemiluminescence reactions associated with an optical transduction can be used to design highly sensitive biosensors.

In 1990, Gautier et al. [113] investigated highly selective and ultra-sensitive biosensors based on luminescent enzyme systems linked to optical transducers. In this way, a fiber optic sensor with immobilized enzymes was designed; the solid-phase bio-reagent was maintained in close contact with the tip of a glass fibre bundle connected to the photomultiplier tube of a luminometer. A bacterial luminescence fiber optic sensor was used for the determination of NADH. Various NAD(P)-dependent enzymes, sorbitol dehydrogenase, alcohol dehydrogenase and malate dehydrogenase, were co-immobilized on preactivated polyamide membranes with the bacterial system and used for the determination of sorbitol, ethanol and oxaloacetate at the nanomolar level with a good precision. The same authors, also in 1990, described a multi-function biosensor for the determination of either ATP or NADH using a single bioluminescence-based fiber optic probe [114]. This was made possible by co-immobilizing the firefly luciferase from Photinus pyralis for ATP analysis with the bacterial luciferase/oxidoreductase system from Vibrio harveyi for NADH analysis, on the same pre-activated polyamide membrane. In 1992, Gautier describes the role of membranes in the design of the main types of biosensors proposed in this time [115].

The reproducible and easy immobilization of receptors on sensor surfaces is a prerequisite for the development of receptor-based fiber optic biosensors. Using a fused silica fiber as the transducer, binding processes of luminescently labeled ligands can be monitored by evanescent wave sensor technology. The vesicle fusion technique was chosen for the immobilization of membrane-bound receptors in order to preserve their binding specificity and activity, by embedding them in an environment similar to a lipid bilayer. The results of initial studies of repetitive cycles of lipid layer deposition and removal, indicating good reproducibility of lipid layer formation on the fiber, are presented by Klee et al. [116]. Using the binding of fluorescently labeled streptavidin to a biotinylated lipid layer as a model system for receptor-ligand interaction, good sensitivity, combined with low nonspecific binding were observed.

The characteristics and performance of biosensors mainly depend on the properties of the bioactive layer associated with the transducer. Two approaches are addressed by Blum et al. [117]: the designs of a) a reagentless fiber-optic biosensor with cosubstrates embedded in the vicinity of the immobilized enzyme; b) a compartmentalized layer with sequential enzymatic reactions for improved signal detection. Concerning the reagentless biosensor, Blum et al. focus on the bacterial bioluminescent system involving two enzymes, oxidoreductase and luciferase, for NADH detection. When associating such a self-contained sensing layer with the transducer, the biosensor can be operated for 1.5 h without reloading. For the studies of compartmentalized enzyme layers, the sequential bienzymatic system lactate oxidase-peroxidase is chosen as a model. The hydrogen peroxide produced by the lactate oxidase reaction serves as a cosubstrate for the chemiluminescence reaction of luminol catalyzed by peroxidase. Compartmentalization of the bioactive layer is obtained by immobilizing the two enzymes separately on different membranes stacked at the sensing tip of the fiber-optic sensor. When using such a design, a 20-fold increase of the sensor response for lactate is obtained compared with a sensor including lactate oxidase and peroxidase randomly coimmobilized on the same membrane. Latter, the same research group developed a fiber optic biosensor for the specific and alternate determination of ATP, ADP and AMP [118]. The sensing layer is arranged by compartmentalizing the tri-enzyme sequence adenylate kinase - creatine kinase - firefly luciferase. The two kinases are covalently co-immobilized on a collagen membrane, whereas firefly luciferase is bound alone on a separate one. For the specific determination of each adenylic nucleotide, three particular reaction media are needed with which flow-injection analysis can be performed in the 2.5-2500 pmol for ATP, 10-2500 pmol for ADP and 25-5000 pmol for AMP linear ranges. When the three nucleotides are present simultaneously in the same sample, the transient inhibition of adenylate-kinase activity by adenosine 5′- monosulphate enables their specific and alternate measurement.

Also, a portable biosensor has been developed to meet the demands of field toxicity analysis [119]. This biosensor consists of three parts, a freeze-dried biosensing strain within a vial, a small light-proof test chamber, and an optic-fiber connected between the sample chamber and a luminometer. Various genetically engineered bioluminescent bacteria were freeze-dried to measure different types of toxicity based upon their modes of action. GC2 (lac::luxCDABE), a constitutively bioluminescent strain, was used to monitor the general toxicity of samples through a decrease in its bioluminescence, while specific toxicity was detected through the use of strains such as DPD2540 (fabA::luxCDABE), TV1061 (grpE::luxCDABE), DPD2794 (recA::luxCDABE), and DPD2511 (katG::luxCDABE). These inducible strains show an increase in bioluminescence under specific stressful conditions, i.e. membrane-, protein-, DNA-, and oxidative-stress, respectively. The toxicity of a sample could be detected by measuring the bioluminescence 30 min after addition to the freeze-dried strains. Using these strains, many different chemicals were tested and characterized. This portable biosensor, with a very simple protocol, can be used for field sample analysis and the monitoring of various water systems on-site.

Recently, biotin was covalently coupled with alginate in an aqueous-phase reaction by means of carbodiimide-mediated activation chemistry to provide a biotin-alginate conjugate for subsequent use in biosensor applications [120]. The synthetic procedure was optimized with respect to pH of the reaction medium (pH 6.0), the degree of uronic acid activation (20%), and the order of addition of the reagents. The new biotin-alginate conjugate was used for the encapsulation of bioluminescent reporter cells into microspheres. A biosensor was prepared by conjugating these biotinylated alginate microspheres to the surface of a streptavidin-coated optical fiber, and the performance of the biosensor was demonstrated in the determination of the antibiotic, mitomycin C as a model toxin.

Also, ionic and colloid gold influence on luminous bacteria Photobacterium phosphorum B7071 bioluminescence have been studied by Gruzina et al. [121]. It was shown that both forms of gold inhibited bioluminescence of the studied bacteria depending on their concentrations and incubation time with cells. The approaches to creation of biosensor systems based on luminous bacteria and semiconductor structures or optical fibers are proposed.

Eltoukhy et al. [122] describes a bioluminescence detection lab-on-chip consisting of a fiber-optic faceplate with immobilized luminescent reporters/probes that is directly coupled to an optical detection and processing system-on-chip fabricated in a 0.18 μm process. The lab-on-chip is customized for such applications as determining gene expression using reporter gene assays, determining intracellular ATP, and sequencing DNA.

On the other hand, hydrophobic phosphorescent Pt-porphyrins have been used for the development of luminescent polymer films designed for fibre-optic oxygen sensors [123]. Luminescent and quenching characteristics of several Pt-porphyrins incorporated into polymer matrices have been studied to optimize the preparation of sensitive coatings for fibre-optical sensors. The films thus obtained have been used for fibre-optical oxygen monitoring in solutions. More recent, a simple system for enzymatic flow-injection analysis of metabolites is described by Ovchinnikov et al. [124], which is based on the phosphorescence lifetime based detection of molecular oxygen using phase-modulation techniques and a simple instrument phosphorescence phase detector equipped with a fibre-optic probe. The phase detector is connected to the oxygen sensor membrane and allows real-time continuous monitoring of the phosphorescence phase shift. This parameter is related to the phosphorescence lifetime of the oxygen probe, therefore giving a measure of the dissolved oxygen concentration, and its changes as a result of the enzymatic oxidative reaction with the substrate. The sensor membrane is positioned in a compact integrated flow-through cell and exposed to the flow stream. Using glucose as a test analyte and glucose oxidase enzyme, two different sensor setups were tested: 1) the membrane type biosensor in which the enzyme is immobilized directly on the oxygen sensor membrane; 2) the microcolumn type biosensor in which the enzyme is immobilized separately, on a microparticle sorbent (controlled pore glass) and put into a microcolumn with the oxygen sensor membrane placed at the column outlet. In either case a new type of oxygen sensitive material was used, which provides a number of advantages over the existing materials. In this material the oxygen-sensitive coating was applied on a microporous scattering support, the latter comprised of a layer of cellulose particles on polyester support.

7. REFRACTIVE INDEX

Akkin et al. [125] describe a fiber-based optical biosensor, which is capable of detecting ultra-small refractive index changes in highly scattering media with high lateral and longitudinal spatial resolution. The system is a dual channel phase-sensitive optical low coherence tomography system that measures relative optical path length differences between the orthogonal modes of the polarization-maintaining fiber.

Tubb et al. [126] describe a new design of optical fibre surface plasma wave chemical sensor. The basic sensor consists of a tapered single-mode optical fibre with a thin layer of silver evaporated onto the tapered section. The gradually changing diameter of the fibre along the taper and the variation in silver depth around the taper result in a distributed coupling between the guided mode of the fibre and the surface plasma wave. As a result, the coupling to the surface plasma wave occurs over an enlarged spectral range. The device shows good sensitivity to refractive index with refractive index changes of 5×10^{-4} being detectable.

Also, a biosensor based on long period grating (LPG) technology has been used to demonstrate the detection of large molecules (proteins) and small molecules (pesticides) [127]. The LPG sensor is a spectral loss optical fiber based system that provides direct detection of large molecules, by using an antigen or antibody modified hydrogel, without the need for secondary amplification. The binding of the specific target results in a mass increase that produces a localized refractive index change around the LPG region and thus a spectral shift in the observed wavelength loss band. The magnitude of the observed shift can be correlated to target concentration. The HIV protein p24 was directly detected at 1 ng/mL with a specific signal that was 5-7 times that of the system noise. A direct and indirect competitive assay was demonstrated with the target atrazine. The sensitivity of the two competitive assay formats was in the range of 10-50 ng/mL.

The investigation group of Cheng presents a novel class of label-free fiber-optic localized surface plasmon resonance sensor which retains many of the desirable features of the propagating surface plasmon resonance sensors, namely, the sensitivity to the refractive indices of bulk liquids and the ability to interrogate biomolecular interactions without a label [128,129]. The sensor was constructed on the basis of modification of the unclad portion of an optical fiber with self-assembled Au colloids. The sensor is easy to fabricate and can be constructed by simple optical designs. Moreover, the sensor has the potential capability for on-site, in vivo, and remote sensing, can be easily multiplexed to enable high-throughout screening of biomolecular interactions, and has the potential use for disposable sensors.

Optical sensors based on the excitation of surface plasmons (SP) have proven to hold great potential for biomolecular interaction analysis and detection of biological analytes. In order to reach out from centralized laboratories, the surface plasmon resonance (SPR) sensors have to be developed into robust portable sensing devices capable of operating in the field. In this way, an optical fiber SPR sensor based on polarization-maintaining fibers and wavelength modulation is presented by Piliarik et al. [130]. It is demonstrated that this design provides superior immunity to deformation of optical fibers of the sensor and, thus allows for more accurate SPR measurements under realistic operation conditions. Experimental results indicate that this fiber-optic SPR sensor is able to resolve refractive index changes as low as 4×10^{-6} under moderate fiber deformations.

The micro- and nano-scale miniaturization of chemical and biochemical sensors is of great scientific and technological interest and in the last decade, much effort was focused on miniaturizing fiber-optic SPR sensors. In this way, a highly sensitive micrometer-sized optical fiber affinity biosensor is reported based on the localized SPR in gold nanoparticles adsorbed at an end-face of an optical fiber; this sensor probes the affinity between biological molecules in real time without any labeling of the analyte; the highest resolution of 10^{-5} in refractive index units is demonstrated with a red-light-emitting diode used as a light source [131]; a novel class of fiber-optic biosensor that exploits the localized SPR of self-assembled gold colloids on the grating portion of a long-period fiber grating is proposed [132]; Chang et al. [133,134] present a sensitive nano-optical fiber biosensor made by shaping a fiber to form a taper with a tip size under 100 nm; a 3-D coded finite-difference time-domain approach verifies the excitation of the surface plasmon wave and the differences among its intensities in media of various refractive indices; Lin et al. [135] developed a side-polished multimode fiber sensor based on SPR as the transducing element with a halogen light source; the SPR fiber sensor is side polished until half the core is closed and coated with a 37 nm gold thin film by dc sputtering; the SPR curve on the optical spectrum is described by an optical spectrum analyzer and can sense a range of widths in wavelengths of SPR effects; the measurement system using the halogen light source is constructed for several realtime detections that are carried out for the measurement of the index liquid detections for the sensitivity analysis; the sensing fiber is demonstrated with a series of refractive index liquids and a new type of the fiber-optic microsensor for SPR was created on the basis of the fabrication technology of optical fiber probes in near-field scanning optical microscopy (NSOM) [136]. The newly developed SPR microsensors were prepared by coating a gold-metallic film on the chemically etched single-mode fiber containing a conical core. They were applied to the real-time monitoring of the refractive index (RI) of transparent liquids flowing in the microfluidic device.

In the last years, high sensitivity chemical and biological sensors based on etched core fiber Bragg gratings that detect change in the index of refraction of surrounding solutions, were developed to measure the index of refraction of different solutions [137-140] and compact three segmented multimode fiber modal interferometer for high sensitivity refractive-index measurement also have been described [141].

On the other hand, fiber-optic waveguides based Micro-Opto-Electro-Mechanical Systems (MOEMS) form a significant class of biosensors which have notable advantages like light weight, low cost and more importantly, the ability to be integrated with bio-systems. Iintegrated microfluidic fiberoptic waveguide biosensor is presented by Chandrasekaran and Packirisamy [142]. The phenomenon of evanescence is employed for sensing mechanism of the device. Herein, the fiber-optic waveguide is integrated with bulk micromachined fluidic channel across which different chemical and biological samples are passed through. The significant refractive index change due to the presence of biological samples that causes the evanescent field condition in the waveguides leads to optical intensity attenuation of the transmitted light. The study of the modulation in optical intensity is used to detect the properties of the species used in the evanescent region. The intensity modulation of light depends upon the geometry of the waveguide, the length of evanescent field, the optical properties of specimen used for producing evanescence and the changes in the properties by their reaction with other specimen. Therefore, this device is proposed for biosensing applications.

8. OTHER TECHNIQUES

Pepper [143] discuss the potential use of nonlinear optical phase conjugation to enhance the performance of various classes of optical interferometric sensors and optical fiber biosensor devices, with potential application to trace-compounds detection of explosives and other species. Examples include Michelson interferometers, laser homodyne sensors, ellipsometers, and modulation spectrometers. Compensated interferometric devices using nonlinear optical phase conjugation may lead to a new class of fieldable remote sensor which is robust, compact, inexpensive, and portable, with the capability of functioning in real-world environments.

An interferometric optical fiber microcantilever beam biosensor has successfully demonstrated real time detection of target molecules [144]. The microcantilever biosensor effectively combines advanced technology from silicon micromachining, optical fiber sensors, and biochemistry to create a novel detection device. This approach utilizes affinity coatings on micromachined cantilever beams to attract target molecules. The presence of the target molecule causes bending in the cantilever beam, which is monitored using an optical displacement system. Dose-response trials have shown measured responses at nanogram ml^{-1} concentrations of target molecules.

A D-type fiber biosensor based on SPR technology and heterodyne interferometry is presented by Chiu et al. [145]. The sensing device is a single-mode optical fiber in which half the core is polished away and a thin-film layer of gold is deposited.

A fiber-optic sensor is designed based on multicavity Fabry-Perot interferometry for the study of optical thickness in self-assembled thin-film layers [146]. This miniature sensor is applicable not only to the measurement of self-assembled polyelectrolyte layers but also to the immobilization of proteins such as immunoglobulin G.

Recently, a new method of optical switch design is proposed, which is used in optical fiber biological protein chips [147].

Also, in the last year, Rindorf et al. [148] present the first incorporation of a microstructured optical fiber (MOF) into biochip applications. A 16-mm-long piece of MOF is incorporated into an optic-fluidic coupler chip, which is fabricated in PMMA polymer using a CO_2 laser. The developed chip configuration allows the continuous control of liquid flow through the MOF and simultaneous optical characterization. While integrated in the chip, the MOF is functionalized towards the capture of a specific single-stranded DNA string by immobilizing a sensing layer on the microstructured internal surfaces of the fiber. The sensing layer contains the DNA string complementary to the target DNA sequence and thus operates through the highly selective DNA hybridization process. Optical detection of the captured DNA was carried out using the evanescent-wave-sensing principle. Owing to the small size of the chip, the presented technique allows for analysis of sample volumes down to 300 nL and the fabrication of miniaturized portable devices.

Such as we described above, there is an increasing demand for sensitive detection systems which are required to detect analytes in often very dilute and diverse circumstances. Fiber-optic biosensors have recently gained a lot of interest, and different schemes have been proposed and applied to detect a wide variety of analytes. The antibody-antigen or enzyme-analyte recognition or reaction provides the basis for the sensitivity and/or selectivity of the reaction. The transduction of the biochemical signal to an electrical signal is often a critical step wherein a large fraction of the "signal loss", for example, fluorescence (by quenching) may occur. This leads to deleterious effects on the sensitivity and selectivity of the biosensor, besides decreasing the quality of the reproducibility of the biosensor. A fractal analysis is presented for the binding of pyrene in solution to β-cyclodextrin attached to a fiberoptic chemical sensor [149]. The fractal analysis provides novel physical insights into the reactions occurring on the fiber-optic chemical surface and should assist in the design of fiber-optic chemical sensors [150,151].

Noto et al. [152] report on molecular weight dependence measurements for an optical resonance biosensor. A dielectric micro-particle is evanescently coupled with an optical fiber for the resonance stimulation, and a shift of the resonance wavelength is measured to monitor protein monolayer formation on the micro-particle surface. Wavelength shifts for proteins over two orders of magnitude in molecular weight are measured.

Walt use optical imaging fibers to fabricate a chemical and biochemical sensor that utilizes the ability of living cells to respond to biologically significant compounds. The sensor is created by randomly dispersing single NIH 3T3 mouse fibroblast cells into an optically addressable fiber-optic microwell array such that each microwell accommodates a single cell. The cells are encoded to identify their location within the array and to correlate changes or manipulations in the local environment to responses of specific cell types [153,154].

The fabrication and testing of a novel waveguide based biosensor for sensing in microchannels is presented by Deverkadra and McShane [155]. Unlike evanescent wave sensing, which has large intrinsic losses, an absorption based sensing scheme was introduced. The fabricated waveguide sensor has excellent transparency in the UV-Vis region of the spectra and is able to perform simultaneous detection of multiple analytes in microchannels. The optical interconnection to the device was achieved using self-aligned V-grooves etched on the silicon wafer using KOH etching. The fabricated device has applications in micro total analysis systems for measuring various analytes in the microchannels, tissue engineering for continuous measurement of the oxygen in the artificial scaffolds, and fluorescence-based measurement by attaching fluorescent dyes to the waveguide end face in the microchannel.

Finally, Konry et al. [156,157] demonstrate that it is possible to create surface-conductive fiber optics, upon which may be electropolymerized a biotinylated polypyrrole thin film, which may then be used to affinity coat the fiber with molecular recognition probes. This fiber-optic electroconductive surface modification is done by the deposition of a thin layer of indium tin oxide. Thereafter, biotin -pyrrole monomers are electropolymerized onto the conductive metal oxide surface and then exposed to avidin. Avidin biotin interactions were used to modify the fiber optics with biotin-conjugated cholera toxin B subunit molecules, for the construction of an immunosensor to detect cholera antitoxin antibodies.

9. APPLICATIONS

Optical fibers have been used for a variety of sensing applications. The small physical dimensions and the ease with which they can be used for multiplexed analyte detection make them ideal platforms for sensing.

Bioanalytical microsystems based on miniaturized biosensing elements could find wide applications in DNA analysis, drug discovery, medical diagnostics, and environmental monitoring as well as in protection against bioterrorism. Miniaturization, portability, multianalyte potential, and interfacing with electronic functions are critical elements for biosensing devices to meet the demands in these fields.

Optical fibers have been used to develop sensors based on nucleic acids and cells. Sensors employing DNA probes have been developed for various genomics applications and microbial pathogen detection. Live cell-based sensors have enabled the monitoring of environmental toxins, and have been used for fundamental studies on populations of individual cells [158].

Advances in nanotechnology have recently led to the development of fiber optics-based nanosensor systems having nanoscale dimensions suitable for intracellular measurements. The possibilities to monitor in vivo processes within living cells could dramatically improve our understanding of cellular function, thereby revolutionizing cell biology. Fiber optic sensors provide significant advantages for in situ monitoring applications due to the optical nature of the excitation and detection modalities. Fiber optics sensors are not affected by electromagnetic interferences from static electricity, strong magnetic fields, or surface potentials. Another advantage of fiber optic sensors is the small size of optical fibers, which allow sensing intracellular/intercellular physiological and biological parameters in microenvironments. Biosensors, which use biological probes coupled to a transducer, have been developed during the last two decades for environmental, industrial, and biomedical diagnostics. Nanotechnology has been revolutionizing important areas in molecular biology, especially diagnostics and therapy at the molecular and cellular level. The combination of nanotechnology, biology, and photonics opens the possibility of detecting and manipulating atoms and molecules using nanodevices, which have the potential for a wide variety of medical uses at the cellular level. The nanoprobes were fabricated with optical fibers pulled down to tips with distal ends having sizes of approximately 30–50 nm. The nanoscale size of this new class of sensors, allows for measurements in the smallest of environments. One such environment that has evoked a great deal of interest is that of individual cells. Using these nanobiosensors, it has become possible to probe individual chemical species in specific locations throughout a cell [159,160].

Biosensors utilizing immobilized antibodies have become increasingly useful due to the specificity of antigen–antibody interactions and the fact that the biosensors can be used where typical cuvette-based spectroscopic measurements are not practical. Of particular usefulness are fiber optic-based biosensors, as the fiber upon which the antibody is immobilized serves as a conduit for measuring optical changes occurring during the antigen–antibody binding event.

Fluorescence-based biosensors provide advantages over other forms of biosensors in that the high signal- to-background discrimination associated with fluorescence is coupled to the antigen binding event. Biosensors utilizing fluorescence may also take advantage of evanescent wave excitation, where excitation light propagating through the fiber excites fluorophores within close proximity (.100 nm) to the fiber surface. Excitation via this technique results in fluorescencefrom labeled antigens only when the antigensbound by antibodies are within the evanescent wave. Thus, background signals from bulk solution contribute little or nothing to the total measured fluorescence, in contrast to cuvette-based measurements where signal from unbound antigen may dominate. Evanescentwave biosensors are also extremely useful for conducting analyses in opaque media, heterogeneous solutions, or colloidal suspensions.

One continuing problem of using fluorescence-based fiber optic biosensors is that of variations in signal response among individual fibers. Wadkins et al. [161] addressed this problem by labeling a portion of the immobilized capture antibody with the fluorescent cyanine dye Cy5.5 (emission λ_{max} = 696 nm). The antigen was then labeled with fluorescent Cy5 (emission λ_{max} = 668 nm). Both fluorophores were excited by 635 nm light, and their emission was collected using both a fiber optic spectrometer and a biosensor optimized to collect fluorescence at two wavelengths. The fluorescence from the Cy5.5-

labeled capture antibody served as a calibration signal for each fiber and corrected for differences in optics, fiber defects, and varying amounts of capture antibody present on the fiber.

Also, Alexa Fluor 647 (AF647) is a relatively new fluorescent dye, which has peak excitation at 650 nm and peak emission at 665 nm. It has a molecular weight of aproximately 1300. These properties are similar to the more commonly employed dye Cy5. On the other hand, the RAPTOR is a portable, automated biosensor that, like its predecessor the Analyte 2000, is useful for on-site analysis of food, water, or clinical samples for biological contaminants. The performance of the AF647 was explored as an alternative to Cy5 for immunoassays on the RAPTOR. The RAPTOR performs sandwich fluoroimmunoassays on the surface of small polystyrene optical waveguides for analyte detection. Primarily, due to the self-quenching characteristics of Cy5, AF647 is substantially more effective in fluoroimmunoassays, yielding over twice the signal for any given analyte concentration. The limitations of Cy5 were elucidated with an immunoassay for ricin, while the advantages of AF647 were demonstrated in both direct binding assays as well as in a sandwich immunoassay for staphylococcal enterotoxin B [162]. The principal applications of these biosensors are summarized in table 2.

10. CONCLUSIONS

Fiber-optic biosensors will play a significant role in the development of biosensors because they can be easily miniaturized and integrated for the determination of different target compounds. These biosensor types have been the objective of a large number of investigations in the last years and they provide numerous ways of performing the rapid, remote, in-line and on-line determination of a lot types of analytes in a wide range of application fields. They are under continuous development and research in this area places increasing emphasis on the works concerning the sensors' performance, such as micro-structural stability, leaching, reversibility, response time, repeatability, sensitivity and selectivity, instead of simply demonstrating the sensing potential. Also, rapid advances have been made in improving immobilization protocols.

Table 2. Applications of fiber optic biosensors.

Analyte	Remarks	Applications	Ref.
Sensitivity of 0.01 nmol l^{-1} obtained from detection of Cy5 serial solutions with various concentrations	Based on principle of evanescent wave while light	Legionella pneumophila	163
Biological agent	Analyte 2000 fiber optic biosensor, which performs four simultaneous fluorescent sandwich immunoassays on the surface of tapered optical probes	Aerosolized bacterial sample	164
Bacillus globigii spores, ovalbumin, Erwinia herbicola, and MS2 coliphage	Employing evanescent wave illumination on polystyrene fiber optic waveguides, RAPTOR performed fluorescent immunoassays	During a 4-day laboratory trial assaying 144 blind samples	165
Giardia cysts (parasitic protozoan)	Using a fluorescent sandwich immunoassay on surface of short polystyrene optical probes with capture antibody adsorbed to probe surface; target analytes bound to fiber by capture antibodies are detected with fluorescently labeled tracer antibodies, which are held in a separate reservoir	Water quality monitoring	166
Escherichia coli O157:H7 and Shiga-like toxins	Polyclonal antibody immobilized on polystyrene fiber waveguides through a biotin-streptavidin reaction that served as the bacteria and toxin capture entity; AF647 dye-labeled antibodies against E. coli O157:H7 or SLTS incubated with the waveguides used to detect cells or toxin and generate a specific fluorescent signal, which was acquired by launching a 635 nm laser-light from an Analyte 2000; fluorescent molecules within several hundred nanometers of the fiber excited by evanescent wave, and a portion of the emission light from fluorescent dye transmitted by the fiber and collected by a photodetector at wavelengths of 670 to 710 nm quantitatively	Ground beef samples	167
Escherichia coli O157:H7	Using antibody-immobilized biconical tapered fiber sensors	First order kinetic model proposed to quantify rate of attachment to pathogen to sensor surface	168
Detection of Escherichia coli O157:h7	With a portable evanescent-wave biosensor with silica and polystyrene waveguides; by launching light from a 635-nm laser diode	In 10 and 25 gram ground beef samples	169
Pathogens	Based upon antigen-antibody interactions; sensor	Food-borne	170

	comprises a nanoporous GeSe channel waveguide fabricated on a substrate, with an intermediate cladding buffer layer ($GeSe_2$), which is required when the substrate does not transmit at the desired λ; light from a laser source is then coupled through a fiber and prism into the waveguide and collected with the help of a lens into a detector	pathogens	
Detection of live Pseudomonas aeruginosa	Using a sensing film containing a fourth-generation hydroxy-terminated polyamidoamine (PAMAM) dendrimer (i.e., G4-OH) and SYTOX Green fluorescent nucleic acid stain	In water and air	171 172 173
Detection of Protein A produced by Staphylococcus aureus	In immunosensor, using a 40 mV argon-ion laser that generated laser light at 488 nm together with plastic optical fiber and antibodies to protein A physically adsorbed onto the fiber; principle of detection involved a sandwich immunoassay with fluorescein isothiocyanate conjugated with anti-(protein A) immunoglobulin G to produce signals of the antigen-antibody reaction	In clinical specimens and foods	174
Staphylococcus aureus and Streptococcus pneumoniae colonies	By evanescent wave fibre optic sensor; biosensor monitors interactions between analytic (bacteria) and the evanescent field of an optical fiber passing through the culture media where the bacteria grows	Detecting and monitoring aerobiological pathogen contamination in hospital environment	175 176
Detection of listeria monocytogenes and salmonella enteritidis	Two optical evanescent wave immunosensors evaluated: Antibody-coupled fiber-optic biosensor and SPR immunosensor; in fiber optic sensor, polyclonal antibodies for test organisms immobilized on polystyrene fiber wave-guide with streptavidin-biotin chemistry	In food	177
Detection of listeria and Salmonella typhimurium	Using principle of fluorescence resonance energy transfer (FRET);labeled antibody-protein G (PG) complexes formed via incubation of anti-Salmonella antibodies labeled with FRET donor fluorophores (AF546) and PG labeled with FRET acceptor fluorophores (AF594); using silanization, labeled antibodies-PG complexes were then immobilized on decladded, tapered silica fiber cores to form evanescent wave-sensing region	In ground pork samples	178 179
Salmonella	Compact fiber-optic immunosensor using tapered fibers and acoustic enhancement	Biomedical applications	180

Monitoring of the genetic responses of all the cells in the array	Cell array biosensor composed of thousands of individual bacteria cells; single cell array produced by immobilizing cells on optical imaging fiber; high-density array of microwells fabricated on optical imaging fiber's distal face by selectively etching individual fibers cores; each microwell used to accommodate a single living bacterium	A reporter gene that responds to the presence of environmental pollutants	181
Detection of Yersinia pestis fraction 1 antigen	Using argon ion laser (514 nm) to launch light into a long-clad fiber and measures fluorescence produced by an immunofluorescent complex formed in the evanescent wave region	In phosphate-buffered saline, serum, plasma, and whole blood	182
Nucleic acids	Fiber optic evanescent wave biosensor by using a 1 mm crystal fiber and a 635 nm laser diode; optical fiber chemically modified to produce a silylated surface; covalent binding occurs applying an amino-terminal nucleic acid to surface of fiber; to test sensitivity and specificity of biosensor, three consecutive trials were performed: direct detection of Cy5-dCTP serial dilution, model hybridization of two oligonucleotide probes and oligonucleotide probe hybridization with the target nucleic acid labeled with Cy5 by PCR amplification	Detection of bacterial DNA	183
Nucleic acids	Based on chemiluminescence; DNA probes covalently immobilized onto distal end of optical fiber bundle; hybridization of HRP (horseradish peroxidase)-labeled complementary nucleotides to the immobilized probes was detected by enhanced chemiluminescence	Detection of DNA hybridization; diagnostic applications	184
Nucleic acids	Short single-stranded DNA (ssDNA) oligonucleotides can be grown on the surface of fused silica by automated nucleic acid synthesis; immobilized ssDNA deposited at a desired average density	Control of selectivity coefficients and relative surface affinities	185
Nucleic acids	Thiazole orange dyes derivatized with ethylene glycol linkers of various lengths, and covalently linked to the 5' end of the oligonucleotides after solid-phase synthesis; labeled oligonucleotides exhibited enhanced fluorescence upon hybridization to complementary DNA sequences at the surfaces of optical fibers, providing for a self-contained labeling strategy	Development of a self-contained DNA biosensor on a fiber optic surface	186 187
Nucleic acids	Surfaces of two sets of fused silica optical fibers	Fluorimetric	188

	were functionalized with hexaethylene oxide linkers from which decaadenylic acid oligonucleotides	detection of T/AT triple-helical DNA formation	
Nucleic acids	Based on hybridization and temperature-induced dissociation of synthetic oligonucleotides.; hybridization pair consists of biotin labeled 38 mer oligonucleotide immobilized to streptavidin-coated optical fiber and fluorescently-labeled near-complementary oligonucleotide reporter sequence	Detection of radiation-induced or chemically-induced oxidative DNA damage	189
Nucleic acids	Based on the combination of photon counting and laser induced fluorescence in the evanescent field on the optical fiber surface	Characterization of nucleic acid hybridization	190
Nucleic acids	Evanescent wave DNA biosensor based on novel molecular beacons (MB) which are oligonucleotide probes that become fluorescent upon hybridization with target DNA/RNA molecules; biotinylated MBs designed and immobilized on an optical fiber core surface via biotin-avidin or biotin- streptavidin interactions	Analysis of specific γ-actin mRNA sequences amplified by polymerase chain reaction	191 192 193
Simultaneous detection of multiple genes	Optical fibers were first treated with poly-l-lysine, and then were made into fiber-optic DNA biosensors by adsorbing and immobilizing the oligonucleotide probe on its end; assembling fiber-optic DNA biosensors in a bundle in which each fiber carried a different DNA probe	Hybridization of fluorescent labeled cDNA of p53 gene, N-ras gene and Rb1 gene to DNA array was monitored by CCD camera	194
Benzo[a]pyrene DNA adduct	Antibody-based submicron fiber biosensor	Monitoring human exposure and health effects associated with polycyclic aromatic compounds	195 196
Adenosine diphosphate (ADP)	Hexokinase, pyruvate kinase, and glucose-6-phosphate dehydrogenase are coimmobilized in a polyurethane membrane and fixed to a commercial fiber-optic probe; ADP is continuously phosphorylated/dephosphorylated in a cycle catalyzed by the kinases; glucose-6-phosphate formed in hexokinase reaction is oxidized by dehydrogenase, leading to the formation of NADH	Biological applications	197
Adenylic nucleotides (ATP, ADP and AMP)	Luciferin incorporation in the structure of acrylic microspheres with subsequent confinement in a	Measurement of ATP, ADP and AMP	198

	polymeric film	with the same probe	
Hemoglobin protein	By fine tailoring mode dispersion and sensitivity of dual-peak LPGs using light-cladding-etching method	In sugar solution	199
Thrombin	Based on DNA aptamers used as receptors; anti-thrombin DNA aptamers immobilized on silica microspheres, placed inside microwells on distal tip on imaging optical fiber, coupled to a modified epifluorescence microscope	Competitive binding assay using a fluorescein-labeled competitor	200
Detection of activity of telomerase	Based on total internal reflection fluorescence measuring activity of telomerase on sensor surface	In tumor cells	201
Monitoring protein binding kinetics	For measuring the association and dissociation rate constants of the reaction between biomolecules; study qualitative and quantitative aspects	Biomolecular recognition in real time	202 203
Protein detection	Use of a fiber optic coupler as a platform for bioassays/biosensing	Medical diagnostics, environmental monitoring, and food safety	204
Proteins	Lipoproteins interact with unmodified optical-fiber surface; immunoglobulin G does not bind to glass surface; upon glass surface modification with silane, immunoglobulin G binds significantly to optical fiber; high-density lipoproteins and low-density lipoproteins show a differential interaction with glass surface, and their interaction was altered upon modification of glass surface with cholesterol	Study the interaction of blood proteins with solid surfaces	205
Protein C (PC)	Immuno-optical fiber biosensor that utilizes PC-specific biomolecules (PC probes) tagged with fluorophores; method involves immobilizing monoclonal antibody against PC (anti-PC) on the surface of an optical fiber; when PC in a sample is adsorbed to the anti-PC on the fiber, it can be reacted with the fluorophore tagged PC-probe	To diagnose PC deficiency, to monitor PC level in blood of PC deficient patients, and to measure PC in other PC-containing samples, such as PC producing animal cell culture broth or transgenic animal milk	206
Protein C	A tapered quartz fiber is enclosed in a glass tube (capacity 300 μl) and monoclonal antibody against PC (anti-PC) is immobilized on the surface of this fiber	Medical and biological applications	207 208
Total protein	Based on the principles of fiber optic evanescent wave spectroscopy; applies a dye-immobilized porous glass coating on a multi-mode optical fiber	Liquid sample	209

Protein substances	Optical immune sensors based on photoluminescence of porous silicon and fiber optics in combination with enhanced chemical luminescence	In air	210
Binding and studying 6×His-tagged proteins	Novel graft copolymer, poly-(L-lysine)-graft-poly(ethylene glycol) (PLL-g-PEG) with part of the PEG chains carrying a terminal nitrilotriacetic acid group was synthesized for polymeric interface	Biological applications	211
Glutamate	Encapsulated GDH in silica sol-gel film on tip of optical fiber; GDH catalyzes oxidative deamination of glutamate to α-ketoglutarate and simultaneous reduction of NAD+ to NADH; To quantify glutamate, observing rate of change of NADH fluorescence as a function of time; important consideration for continuous in vivo monitoring is incorporation of a self-sustaining NAD+ source	Photochemical enzyme co-factor regeneration for continuous glutamate monitoring with a sol-gel optical biosensor	212
L-glutamate	Comprised of an L-glutamate-sensitive fluorescent gel, spin-coated onto the tip of an optical imaging fiber	From the foregut plexus of the Lepidopteran, Manduca sexta	31
Two fluorophores termed the donor and the acceptor	Using the distance-dependent chemical transduction method of FRET	Detection of myocardial infarction	213 214
Detect porcine reproductive and respiratory syndrome virus	Using gold nanoparticles and FRET	Medical diagnostics	215
Detection of antibodies directed against antigens of Ebola virus strains Zaire and Sudan	Using photo immobilization methodology based on a photoactivatable electrogenerated poly(pyrrole-benzophenone) film deposited upon an indium tin oxide modified conductive surface fiber optic	In animal and human sera	216
Detection of antibodies to viral antigen	Based on a poly(pyrrole-benzophenone) film	For the diagnosis of hepatitis C virus	217
Ammonia, urea, urease and IgG	Using Brilliant Yellow in thin cellulose acetate membrane as pH sensor; by use of gas permeable membrane, urease-immobilized membrane and IgG immobilized membrane	Physiological and clinical applications	218
Bilirubin	By an enzyme-based fiber-optic fluorescence biosensor; bilirubin is oxidized to biliverdin by bilirubin oxidase, and determined by measuring fluorescence intensity change caused by oxygen consumption in enzymatic reaction; bilirubin oxidase is immobilized onto an affinity membrane which is	Biological applications	219

	layered onto an optical indicating membrane		
Bilirubin	Detected indirectly, based on fluorescence quenching of tris(4,7-diphenyl-1,10-phenanthroline) ruthenium chloride by dissolved molecular oxygen	In serum	220
Bilirubin	Following the variation of its yellow absorption band in presence of biocatalytic system haemoglobin/glucose oxidase/glucose	Biological applications	221
Bilirubin and other pigments in bile	Measuring blue light absorbance in the ducts; optical sensing system could be incorporated into a manometry catheter to aid sphincter of Oddi manometry shortening procedure time and reduce radiography exposure	Discriminating between biliary and pancreatic ductal systems	222
A wide variety of both high and low molecular mass analytes	Fiber optic evanescent wave immunosensors	Medical diagnostics	223
Colchicine	Using fluorescent dye N-vinylcarbazole, which contains a vinyl unit making it photo-copolymerizable in the presence of a monomer, 1,2-cyclohexanediol diacrylate, on a quartz surface of an optode slide. Prior to the photopolymerization process the quartz surface was modified with γ-(methacryloxy)propyl trimethoxysilane to introduce vinyl units	In pharmaceutical tablets	224
Cholesterol	Silicone-entrapped tris(4,7-diphenyl-1,10-phenanthroline) ruthenium(II) complex, the luminescence of which is sensitive to oxygen changes, is used as an optical transducer of the oxidation of cholesterol by cholesterol oxidase	Serum samples	225
Cholesterol	Based on fluorescence quenching; $Ru(phen)_3^{2+}$ used as fluorescent reagent to prepare fluorescent membrane, and cellulose acetate used as carrier of immobilized cholesterol oxidase; dissolved oxygen and cholesterol content in cholesterol solution were detected by means of phase-shifting and phase lock means	Biological applications	226
Choline-containing phospholipids	Phospholipids are hydrolyzed by the enzyme phospholipase-D to choline which is analyzed with a fiber-optic biosensor based on choline oxidase and oxygen transduction; optimum biosensing layer consists of enzyme immobilized on a nylon membrane and placed onto an oxygen sensitive layer formed by a luminescent Ru(II) complex, tris(4,7-diphenyl-1,10-	In serum	227

	phenanthroline)ruthenium(II), dispersed in a silicone membrane		
Choline	Based on luminol electrochemiluminescence integrated in a flow injection analysis; choline oxidase immobilised either covalently on polyamide (ABC type) or on UltraBind preactivated membranes, or by physical entrapment in a photo-cross-linkable poly(vinyl alcohol) polymer alone or after absorption on a weak anion exchanger, diethylaminoethyl Sepharose	Biological applications	228
Glucose	Using cellulose acetate membrane containing glucose oxidase, 2,7-diaminofluorene dihydrochloride, and sodium N-(3-sulfopropyl)-3,3',5,5'-tetramethylbenzidine; reflectance changes at 580 nm	Biochemical analysis	229
Glucose	Based on oxygen optrode, which measures the consumption of oxygen via dynamic quenching of fluorescence of indicator by molecular oxygen; Glucose oxidase (GOD) is immobilised onto surface of this oxygen optrode by adsorption to carbon black and by crosslinking with glutardialdehyde	Continuous determination of glucose	230
Glucose	Optical fiber oxygen sensor, based on the dynamic quenching of the luminescence of tris(1,10-phenanthroline)-ruthenium(II) cation by molecular oxygen, complex is adsorbed onto silica gel, incorporated in a silicone matrix possessing a high oxygen permeability, and placed at the tip of the optical fiber; enzyme immobilized on surface of oxygen optrode	Measurements have been performed in a flow-through cell in air-equlibrated glucose standard solutions of pH 7.0	231
Glucose	Use conconavalin A as a specific receptor	For sugars and carbohydrates	232
Glucose	Based on electrostatic self-assembly; end face of multimode fiber coated with poly (allylamine hydrochloride) in combination with Prussian blue and enzyme glucose oxidase	Biological applications	233
Glucose	With needle-type hollow container (18-gauge needle), immobilized enzyme membrane and optic fiber probe with ruthenium complex; enzyme membrane prepared from glucose oxidase, azide-unit pendant water-soluble photopolymer and ultra-thin dialysis membrane; optic fiber probe inserted into rolled enzyme membrane placed in needle-type hollow	In fish blood	234

	container		
Glucose	Glucose oxidase is immobilized on cellulose acetate membranes through covalent - cross linking process with glutaraldehyde and Bovine serum albumin	Biological applications	235
Glucose	Using adsorption LB films	In blood	236
Glucose	By means of fluorescence emission spectra of soluble and insoluble glucose oxidase	Biological applications	237
Glucose	Hybrid sensor: two oxygen optodes wherein one optode contains immobilized glucose oxidase	In subcutaneous tissue	238
Glucose	Based on enzymatic reaction of glucose oxidase that catalyzes oxidation of glucose to gluconic acid and hydrogen peroxide while consuming oxygen; Tris(1,10-phenanthroline)ruthenium chloride, an oxygen indicator, used as transducer; ruthenium complex and glucose oxidase incorporated into acrylamide polymer that is attached covalently to a silanized optical fiber tip surface by photocontrolled polymerization	Biological applications	239 240
Glucose	Three different combinations of oxygen transducer and sol-gel immobilized GOx: a) GOx sandwiched between sol-gel layer doped with Ru(dpp) and second sol-gel layer composed of pure sol-gel, b) sol-gel layer doped with Ru(dpp) covered with sol-gel entrapped GOx, c) both GOx and sol-gel powder containing GOx incorporated into single sol-gel phase	Biological applications	241
Glucose	Sol-gel derived ceramic-carbon composite electrode for fabrication of biosensor; based on luminol electrochemiluminescence; electrode consists of graphite powder impregnated with glucose oxidase in a silicate network. In this configuration, the immobilized enzyme oxidizes glucose to liberate hydrogen peroxide and graphite powder provides percolation conductivity for triggering the ECL between luminol and the liberated hydrogen peroxide	Biological applications	242
Glucose and oxygen	Oxygen: Dichlorotris(1,10-phenantroline)-ruthenium(II) hydrate incorporated into an adhesive inorganic-organic hybrid polymer coating (ORMOCER®); Ruthenium/ ORMOCER® layer used with optical fibres to form extrinsic or intrinsic sensor. Glucose: fluorescence is quenched by oxygen	Biological applications	243

	depletion within the layer, which can be linked to glucose by incorporation of suitable enzymes		
Hydrogen peroxide and L-lactate	Optode uses the H_2O_2 dependent oxidation of homovanillic acid by horseradish peroxidase (HRP) as the sensing reaction: HRP was immobilized on a membrane and combined with a bifurcated fiber optic probe; for lactate, use a lactate oxidase-HRP membrane	Biological applications	244
Lactate	Based on bacterial cytoplasmic membranes as the biological recognition element and an oxygen sensitive dye layer as the transducer: CPMs from bacteria with induced lactate oxidase system adsorbed onto a cellulose disk	Detection of lactate with no interference from glucose, fructose or glutamic acid	26
Detection of the lactate dehydrogenase	Fiber optic evanescent wave immunosensor; polyclonal antibody preparation was used in both one-step and two-step assays	Clinical diagnosis	245
Detection of fibrinolytic products	Sol-gel-based biosensor; constructed to be selective toward D dimer antigens, which form from dissolution of cross-linked fibrin clots	In stroke treatment	246
Penicillin	Based on co-immobilization of a pH indicator, fluorescein isothiocyanate and penicillinase on a preactivated biodyne B membrane attached to the end of a bifurcated optical fiber; in conjunction with a flow injection analysis	In some pharmaceutical samples	247
Penicillin	Based on the enzyme penicillinase; fabricated by selective photodeposition of analyte-sensitive polymer matrices on optical imaging fibers	To quantify penicillin produced during a Penicillium chrysogenum fermentation	248 249
Pyruvate	Dual-enzyme fiber-optic biosensor: enzyme layer composed of lactate oxidase and lactate dehydrogenase fabricated at tip of sensing optrode; biosensor based on luminescence detection of consumed reduced nicotinamide adenine dinucleotide and measured fluorescence intensity is related to the bulk concentration of pyruvate	Biological applications	250
Analysis of different amino acids	Biosensor works according to the measurement principle of flow injection analysis; consists of a fluorosensor and a measurement cell operating under pressure; In measurement cell, enzymes and molecular weight enlarged coenzyme PEG (MW 20000)-N6-(2-	Biological applications	251

	aminoethyl)-NAD(H) are confined behind a solid ultra-filtration membrane		
Continuous monitoring of adriamycin in vivo	First, an accurate optical design was used to enhance the intensity of light from a 100 μm optic fiber so the fluorescence signal can be detected. Second, new sol-gel method was used to fix the fluorescence substance 4-(N,N-dioctyl)amino-7-nitrobenz-2-oxa-1,3-diazole (D-70) on the tip of the fiber	Monitoring blood concentrations of drugs and metabilities in the biomedical field	252
Anesthetics and other lipid-soluble compounds	Monitors anesthetic-induced changes in fluorescence emission by labeled liposomes entrapped in a hydrogel at the end of two optical fibers	In gas, liquid or solid phases (e.g., blood and tissue)	253
Cocaine metabolites	Analyte 2000, a four-channel fiber optic biosensor using a competitive fluorescence immunoassay	In urine	254
Cocaine	Monoclonal antibody made against benzoylecgonine, a major metabolite of cocaine, was immobilized covalently on quartz fibers and used as the biological sensing element in the portable fluorometer. Benzoylecgonine-fluorescein (BE-FL) was used as the optical signal generator when it bound to the fiber	In coca leaf extracts	255
Ionic surfactants	Consists of a silanized silica optical fiber, onto which acrylodan-labeled bovine serum albumin is immobilized	Biological applications	256
Detection of toxins and other large molecules	Capture antibodies immobilized on the exposed core, which has been tapered for improved sensitivity	Clinical and environmental analyses	257
Detection of staphylococcal enterotoxin B	Fiber optic surface plasmon resonance (SPR) biosensor: based on spectral interrogation of surface plasmons in a miniature sensing element based on a side-polished single-mode optical fiber with a thin metal overlayer. For specific detection of SEB, the SPR sensor is functionalized with a covalently crosslinked double-layer of antibodies against SEB	Biological applications	258 259
Mycotoxin fumonisin B1 (FB1)	Monoclonal antibodies produced against FB1 covalently bound through a heterobifunctional silane to an etched 800 μm core optical fiber. An evanescent wave effect was utilized to excite fluorescein isothiocyanate labeled FB1 (FB1-FITC) molecules near the surface of the fiber	Biological applications	260
Mycotoxins: fumonisins and aflatoxins	Two formats, competitive and non-competitive: a) Fumonisin monoclonal antibodies were covalently	In maize	261

	coupled to an optical fiber and the competition between FB_1 and FB_1 labeled with fluorescein for the limited number of binding sites on the fiber was assessed; b) mycotoxin aflatoxin B_1 was a non-competitive assay using the native fluorescence of this mycotoxin		
Clostridium botulinum toxin A	Evanescent wave of tapered optical fiber for signal discrimination; 50 mW argon-ion laser, which generates laser light at 514 nm, used in conjunction with optical fiber probe that is tapered at distal end; antibodies specific for C. botulinum covalently attached to surface of tapered fiber; sandwich immunoassay using rhodamine-labeled polyclonal anti-toxin A immunoglobin G antibodies for generation of specific fluorescent signal	Biological applications	262
Genotoxicants	Whole-cell bacterial sensors genetically engineered to react to target toxicants by the induction of a selected promoter and the subsequent production of bioluminescent light through a recombinant lux reporter	Environmental monitoring	263
Ricin	Evanescent wave fiber-optic biosensor; sandwich immunoassay scheme; anti-ricin IgG immobilized onto surface of optical fiber in two different ways: a) antibody directly coated to silanized fiber using a crosslinker, b) avidin-coated fibers incubated with biotinylated anti-ricin IgG to immobilize antibody using an avidin- biotin bridge	In buffer solution and river water	264
Direct determination of organophosphate nerve agents	Using recombinant Escherichia coli with surface-expressed organophosphorus hydrolase	On-line monitoring of the detoxification process for organophosphate pesticides-contaminated wastewaters	265 266
Polychlorinated biphenyls	Quartz fiber coated with partially purified polyclonal anti-PCB antibodies; optical signal generated by binding and subsequent fluorescence of the fluorescein conjugate of 2,4,5-trichlorophenoxybutyrate to Ab-coated fiber	Environmental monitoring	267
Detection of imazethapyr herbicide	Purified sheep antibody immobilized on quartz fibers; mixture of fluorescein-labelled imazethapyr analog and free imazethapyr was presented to the fiber	Environmental monitoring	268

	for direct competition of the antibody binding sites or displacement of a previously fluorescein labelled fiber		
Toxic compounds: some herbicides	Based on kinetic measurements of chlorophylla fluorescence in Chlorella vulgaris cells; microalgae immobilized on removable membranes placed in front of tip of an optical fiber bundle inside a homemade microcell	Measure concentration of a toxic chemical in the form of a single drop or dissolved in a continuous flow	269
Detection of photosynthetic herbicides	Based on use of reaction centre isolated from Rhodobacter sphaeroides, a purple bacteria; by time-resolved absorption	In water	270
Herbicides	Optical whole-cell biosensor using Chlorella vulgaris	Environmental monitoring	271
Detection of methsulfuron methyl	Based on competitive immunoreactions between coating-haptens and free haptens in solutions with corresponding antibodies; ovalbumin-methsulfuron-methyl conjugate immobilized on a microscope slide. Horseradish peroxidase labeled goat anti-rabbit IgG used to generate an optical signal	Environmental monitoring	272
Pesticides propoxur (Baygon®) and carbaryl	The biorecognition element is covalently immobilized onto controlled pore glass beads (CPG) and packed in a thermostatized bioreactor connected to a flow-through cell that contains CPG- immobilized chlorophenol red placed at the common end of a bifurcated fiber optic bundle	In spiked vegetables (onion and lettuce) using ultrasound extraction	273
Detection of organophosphorous-based pesticides	Using chemiluminescence and molecular self-assembly	Environmental monitoring	274
Organophosphorous compounds	Biosensor consisting of acetylcholinesterase/viologen hetero Langmuir-Blodgett film	In contamined waters	275
Captan	Biosensor consisting of a glutathione-S-transferase-immobilized gel film	In contamined waters	276
Methyl parathion	Using Flavobacterium sp. whole cells adsorbed on glass fiber filters as disposable biocomponent	Environmental monitoring	277
Pesticides	Functionalized biosensing surfaces were developed for chemiluminescent immunoassay	In vitro pesticide residual analysis	278
Liquid pollutants	On the surface of a planar substrate composing single-mode channel waveguides	Environmental monitoring	279
Environmental pollutants	Based on immobilized living algae cells; measuring principle was determination of chlorophyll	Environmental monitoring in laboratory	280

	fluorescence depending on the load of water probes with toxic compounds using fiber-optic-based electronic equipment; micro-organisms used Scenesdesmus subspicatus immobilized on filter paper and covered with alginate	and field investigations	
Monitoring waste water pollutants	Two algal whole cells biosensors; both optical and conductometric biosensors based on inhibition of algal alkaline phosphatase and esterase activities. Chlorella vulgaris cells are immobilised on a membrane placed in front of an optical fiber bundle for optical sensing or deposited on the surface of an electrode for conductometric sensing	Environmental monitoring	281
Detection of environmental pollutants and monitoring of bacteria for bioremediation	Two different types: a) continuous flow immunosensor based on displacement of fluorescently-labeled antigen from antibodies immobilized on beads, antigen injected into flow stream that passes over a 100 μL bed volume of antibody-coated beads saturated with fluorescently-labeled antigen, displacement of labeled antigen causes increase in fluorescence, proportional to antigen concentration, to be observed downstream; b) fiber optic biosensor, utilizes long, partially clad optical fibers; antibodies immobilized onto fiber core in unclad region at distal end of fiber. Upon binding of antigen and a fluorescent molecule, a change in fluorescence signal is observed	Environmental monitoring	282
2,4,6-Trinitrotoluene	Using a portable fiber optic biosensor; larger "breadboard" system capable of monitoring a single probe was replaced by a lightweight, portable sensor that can monitor 4 optical probes simultaneously	Transition from laboratory to on-site environmental monitoring	283
Small molecule analytes	Fieldable environmental monitor based on antibodies, fiber optic probes, and compact optoelectronics; antibodies immobilized to surface of etched optical fiber tips; displacement reaction between target molecules and fluor-labeled target analogs monitored remotely	In groundwater	284
Determination of low biochemical oxygen demand (BOD)	Artificial wastewater solution employed as standards for calibration of the BOD sensor	In river waters	285
BOD	Sensing membrane at tip of fiber consists of layers of (a) an oxygen-sensitive fluorescent material, (b) Trichosporoa cutaneum immobilized in poly(vinyl	Environmental monitoring	286

	alcohol), and (c) a substrate-permeable polycarbonate membrane to retain the yeast cells. The layers are placed, in this order, on an optically transparent gas-impermeable polyester support. Tris(4,7-diphenyl-1,10-phenanthroline)ruthenium(II) perchlorate is used as the oxygen indicator		
BOD	Based on different microorganisms immobilized in ormosil matrixes	Environmental monitoring	287 288
Heavy metals	From inhibition of alkaline phosphatase present on the external membrane of Chlorella vulgaris microalgae; microalgal cells immobilized on removable membranes placed in front of the tip of an optical fiber bundle inside a homemade microcell	Environmental monitoring	289
Heavy metals	Using whole-cell luminescent bacterial sensors in suspension or immobilized onto fibre-optic tips	In EILATox-Oregon samples	290
Heavy metals	Based on the inhibition of urease activity, where the urease is immobilised on ultrabind membrane	Environmental monitoring	291
Heavy metals	Based on immobilized engineered microorganisms and bioluminescence measurements; A strain of microorganisms from Alcaligenes eutrophus (AE1239) was genetically engineered by inserting a luxCDABE operon from Vibrio fischeri under control of a copper-induced promoter	Environmental monitoring	292
Heavy metals	Highly sensitive and selective catalytic DNA biosensors	Environmental monitoring	293
Copper	Protein molecule, site-specifically labeled with a fluorophore that is attached to the distal end of an optical fiber	In sea water	294
Nitric oxide	Based on a dye-labeled heme domain of soluble guanylate cyclase	Cellular applications	295
Nitric oxide	Gas sensor with nanoporous structure prepared by uniformly immobilizing NO_2 sensing reagents at the pores during the sol-gel process	In air samples	296
Trichloroethene and tetrachloroethene	Based on monooxygenases and sol-gel entrapped fluoresceinamine	water samples, without sample preparation step	297
Trinitrotoluene (TNT)	Trinitrobenzenesulfonic acid labeled with a fluorophore and used as the analyte competitor; evanescent wave fiber-optic biosensor	In water	298 299 300 301
pH	Optical fibre pH sensor with an MPC (2-	Physiological	302
	Methacryloyloxyethyl Phosphorylcholine) membrane has been produced and the in vitro properties of this sensor were assessed in the buffer solutions, both with and without protein	applications	

Likewise, one of the major advantages of using optical biosensors in conjunction with optical fibers is that it permits sample analysis to be done over long distances and this has important implications for field monitoring. However, the main drawback apart from being relatively expensive is that optical fibers may suffer from miniaturization problems. Notwithstanding, the application of optical fiber-based nanosensors has become an area of significant interest and various methods have been developed to alleviate the problems arising from miniaturization. In this sense, optical fiber SPR probes present the highest level of miniaturization of SPR devices, allowing for chemical and biological sensing in inaccessible locations where the mechanical flexibility and the ability to transmit optical signals over a long distance make the use of optical fibers very attractive.

In summary, fiber-optic biosensors will play a significant role in the development of biosensors because they can be easily miniaturized and integrated for the determination of different target compounds in a wide variety of application fields, such as industrial process and environmental monitoring, food processing, and clinical applications.

REFERENCES AND NOTES

1. Wolthuis, R.; McCrae, D.; Saaski, E.; Hartl, J.; Mitchell, G. Development of a medical fiberoptic ph sensor based on optical absorption. IEEE Transactions on Biomedical Engineering 1992, 39, 531–537.
2. Wolthuis, R. A.; McCrae, D.; Hartl, J. C.; Saaski, E.; Mitchell, G. L.; Garcin, K.; Willard, R. Development of a medical fiber-optic oxygen sensor based on optical absorption change. IEEE Transactions on Biomedical Engineering 1992, 39, 185–193.
3. Li, Y.; Fan, Y.; Zhang, L.; Ma, S.; Zheng, J. Molecular recognition of functionalized polydiacetylene and its biosensor. Proceedings of SPIE - The International Society for Optical Engineering; 2000; 4077, pp. 242–245.

4. Piletsky, S. A.; Panasyuk, T. L.; Piletskaya, E. V.; Sergeeva, T. A.; El'skaya, A. V.; Pringsheim, E.; Wolfbeis, O. S. Polyaniline-coated microtiter plates for use in longwave optical bioassays. Fresenius' Journal of Analytical Chemistry 2000, 366, 807–810.
5. Llobera, A.; Wilke, R.; Buttgenbach, S. Optimization of poly(dimethylsiloxane) hollow prisms for optical sensing. Lab on a Chip - Miniaturisation for Chemistry and Biology 2005, 5, 506–511.
6. Esteban, O.; Gonzalez-Cano, A.; Diaz-Herrera, N.; Navarrete, M. Absorption as a selective mechanism in surface plasmon resonance fiber optic sensors. Optics Letters 2006, 31, 3089–3091.
7. Bindig, U.; Meinke, M.; Gersonde, I.; Spector, O.; Vasserman, I.; Katzir, A.; Muller, G. IR-Biosensor: flat silver halide fiber for bio-medical sensing? Sensors and Actuators, B: Chemical 2001, 74, 37–46.
8. Rochemont, L. P.; Downer, N. W.; May, T. E.; Smith, H. G.; Oakes, C. E.; Ertan-Lamontagne, M. Evanescent-Wave infrared fiber optic biosensor. Proceedings of SPIE - The International Society for Optical Engineering; 1993; 1796, pp. 349–359.
9. Taga, K.; Kellner, R.; Kainz, U.; Sleytr, U. B. In situ attenuated total reflectance ft-ir analysis of an enzyme-modified mid-infrared fiber surface using crystalline bacterial surface proteins. Analytical Chemistry 1994, 66, 35–39.
10. Lucas, P.; Solis, M. A.; Coq, D. L.; Juncker, C.; Riley, M. R.; Collier, J.; Boesewetter, D. E.; Boussard-Plédel, C.; Bureau, B. Infrared biosensors using hydrophobic chalcogenide fibers sensitized with live cells. Sensors and Actuators, B: Chemical 2006, 119, 355–362.
11. Crespi, F.; Donini, M.; Bandera, A.; Congestri, F.; Formenti, F.; Sonntag, V.; Heidbreder, C.; Rovati, L. Near-Infrared oxymeter biosensor prototype for non-invasive in vivo analysis of rat brain oxygenation: Effects of drugs of abuse. Journal of Optics A: Pure and Applied Optics 2006, 8, S528–S534.
12. Naughton, D. P.; Grootveld, M.; Blake, B. R.; Guestrin, H. R.; Narayanaswamy, R. An Optical Hydroxyl Radical Sensor. Biosensors and Bioelectronics 1993, 8, 325–329.
13. Dyr, J. E.; Tichý, I.; Jiroušková, M.; Tobiška, P.; Slavík, R.; Homola, J.; Brynda, E.; Houska, M.; Suttnar, J. Molecular arrangement of

adsorbed fibrinogen molecules characterized by specific monoclonal antibodies and a surface plasmon resonance sensor. Sensors and Actuators, B: Chemical 1998, 51, 268–272.

14. Dyr, J. E.; Ryšavá, J.; Suttnar, J.; Homola, J.; Tobiška, P. Optical sensing of the initial stages in the growth and development of fibrin clot. Sensors and Actuators, B: Chemical 2001, 74, 69–73.
15. Cunningham, B.; Li, P.; Lin, B.; Pepper, J. Colorimetric resonant reflection as a direct biochemical assay technique. Sensors and Actuators, B: Chemical 2002, 81, 316–328.
16. Yang, Y.; Tan, Y.; Liu, B.; Sun, Y.; Chen, D.; Tan, H. Study of optical fiber biosensor based on white-light interferometry. Hsi-An Chiao Tung Ta Hsueh/Journal of Xi' an Jiaotong University 2003, 37, 914–916.
17. Yang, Y.; Xiong, K.; Tan, Y.; Liu, B.; Sun, Y.; Chen, D.; Tan, H. Direct monitoring of antigen-antibody interactions by optical fiber bioprobe. Proceedings of SPIE - The International Society for Optical Engineering; 2003; 5254, pp. 431–436.
18. Yang, Y.; Wang, Y.; Xiong, K.; Liu, B.; Tan, Y.; Tan, H.; Chen, D. Study of the key technology of optical fiber biosensor. Guangzi Xuebao/Acta Photonica Sinica 2005, 34, 121–125.
19. Homola, J.; Yee, S.S.; Gauglitz, G. Surface plasmon resonance sensors: Review. Sensor and Actuators, B: Chemical 1999, 54, 3–15.
20. Watanabe, M.; Kajikawa, K. An optical fiber biosensor based on anomalous reflection of gold. Sensors and Actuators, B: Chemical 2003, 89, 126–130.
21. Kajikawa, K.; Mitsui, K. Optical fiber biosensor based on localized surface plasmon resonance in gold nanoparticles. Proceedings of SPIE - The International Society for Optical Engineering; 2004; 5593, pp. 494–501.
22. Misiakos, K.; Kakabakos, S. E.; Petrou, P. S.; Ruf, H. H. A monolithic silicon optoelectronic transducer as a real-time affinity biosensor. Analytical Chemistry 2004, 76, 1366–1373.
23. Yan, S.; Mingming, L.; Hong, Z.; Xiao, Y. Y.; Lu, Z. Optical fiber direct-sensing biosensor applied in detecting biolayer thickness of nanometer grade. Proceedings of SPIE - The International Society for Optical Engineering; 2006; 6150 II.

24. Scheper, T.; Buckmann, A. F. A fiber optic biosensor based on fluorometric detection using confined macromolecular nicotinamide adenine dinucleotide derivatives. Biosensors and Bioelectronics 1990, 5, 125–135.
25. Schelp, C.; Sheper, T.; Buckmann, F.; Reardon, K.F. Two fibre-optic sensors with confined enzymes and coenzymes: development and application. Analytica Chimica Acta 1991, 255, 223–229.
26. Ignatov, S. G.; Ferguson, J. A.; Walt, D. R. A fiber-optic lactate sensor based on bacterial cytoplasmic membranes. Biosensors and Bioelectronics 2001, 16, 109–113.
27. Sheper, T.; Brandes, W.; Graw, C.; Hundeck, H.G.; Reinhardt, B.; Ruther, F.; Plotz, F.; Shelp, C.; Schugerl, K.; Schenider, K.H.; Giffhorn, F.; Rehr, B.; Sahm, H. Applications of biosensor systems for bioprocess monitoring. Analytica Chimica Acta 1991, 249, 25–34.
28. Busch, M.; Gutberlet, F.; Hoebel, W.; Polster, J.; Schmidt, H.L.; Schwenk, M. Application of optrodes in FIA-based fermentation process control using the software package FIACRE. Sensors and Actuators, B: Chemical 1993, B11, 407–412.
29. Wang, A.J.; Arnold, M.A. Dual-enzyme fiber-optic biosensor for glutamate based on reduced nicotinamide adenine dinucleotide luminescence. Analytical Chemistry 1992, 64, 1051–1055.
30. Cordek, J.; Wang, X.; Tan, W. Direct immobilization of glutamate dehydrogenase on optical on optical fiber probes for ultrasensitive glutamate detection. Analytical Chemistry 1999, 71, 1529–1533.
31. Issberner, J. P.; Schauer, C. L.; Trimmer, B. A.; Walt, D. R. Combined imaging and chemical sensing of l-glutamate release from the foregut plexus of the lepidopteran, manduca sexta. Journal of Neuroscience Methods 2002, 120, 1–10.
32. Anders, K. D.; Wehnert, G.; Thordsen, O.; Scheper, T.; Rehr, B.; Sahm, H. Biotechnological applications of fiber-optic sensing: Multiple uses of a fiber-optic fluorimeter. Sensors and Actuators, B: Chemical 1993, B11, 395–403.
33. Scheper, T. A fluorometric fiber-optic biosensor for dual analysis of glucose and fructose using glucose-fructose-oxidoreductase isolated from Zymomonas mobilis. Journal of Biotechnology 1994, 36, 39–44.

34. Rogers, K. R.; Cao, C. J.; Valdes, J. J.; Eldefrawi, A. T.; Eldefrawi, M. E. Acetylcholinesterase fiber-optic biosensor for detection of anticholinesterases. Fundamental and Applied Toxicology 1991, 16, 810–820.
35. Rogers, K. R.; Eldefrawi, M. E.; Menking, D. E.; Thompson, R. G.; Valdes, J. J. Pharmacological specificity of a nicotinic acetylcholine receptor optical sensor. Biosensors and Bioelectronics 1991, 6, 507–516.
36. Rogers, K. R.; Valdes, J. J.; Eldefrawi, M. E. Effects of receptor concentration, media ph and storage on nicotinic receptor-transmitted signal in a fiber-optic biosensor. Biosensors and Bioelectronics 1991, 6, 1–8.
37. Mitsubayashi, K.; Kon, T.; Hashimoto, Y. Optical Bio-Sniffer for ethanol vapor using an oxygen-sensitive optical fiber. Biosensors and Bioelectronics 2003, 19, 193–198.
38. Meadows, D. L.; Schultz, J. S. Design, manufacture and characterization of an optical fiber glucose affinity sensor based on an homogeneous fluorescence energy transfer assay system. Analytica Chimica Acta 1993, 280, 21–30.
39. McCurley, M. F. An optical biosensor using a fluorescent, swelling sensing element. Biosensors and Bioelectronics 1994, 9, 527–533.
40. Rex Asties, J.; Miller, W. G. Measurement of free phenytoin in blood with a self-contained fiberoptic immunosensor. Analytical Chemistry 1994, 66, 1675–1682.
41. Ferguson, J. A.; Boles, T. C.; Adams, C. P.; Walt, D. R. A fiber-optic DNA biosensor microarray for the analysis of gene expression. Nature Biotechnology 1996, 14, 1681–1684.
42. Furch, M.; Ueberfeld, J.; Hartmann, A.; Bock, D.; Seeger, S. Ultrathin oligonucleotide layers for fluorescence-based DNA sensors. Proceedings of SPIE - The International Society for Optical Engineering; 1996; 2928, pp. 220–226.
43. Abel, A. P.; Weller, M. G.; Duveneck, G. L.; Ehrat, M.; Widmer, H. M. Fiber-optic evanescent wave biosensor for the detection of oligonucleotides. Analytical Chemistry 1996, 68, 2905–2912.

44. Henke, L.; Piunno, P. A. E.; McClure, A. C.; Krull, U. J. Covalent immobilization of single-stranded DNA onto optical fibers using various linkers. Analytica Chimica Acta 1997, 344, 201–213.
45. Kleinjung, F.; Bier, F. F.; Warsinke, A.; Scheller, F. W. Fibre-Optic genosensor for specific determination of femtomolar DNA oligomers. Analytica Chimica Acta 1997, 350, 51–58.
46. Almadidy, A.; Watterson, J.; Piunno, P. A. E.; Foulds, I. V.; Horgen, P. A.; Krull, U. A fibre-optic biosensor for detection of microbial contamination. Canadian Journal of Chemistry 2003, 81, 339–349.
47. Biran, I.; Rissin, D. M.; Ron, E. Z.; Walt, D. R. Optical imaging fiber-based live bacterial cell array biosensor. Analytical Biochemistry 2003, 315, 106–113.
48. Zeng, J.; Almadidy, A.; Watterson, J.; Krull, U. J. Interfacial hybridization kinetics of oligonucleotides immobilized onto fused silica surfaces. Sensors and Actuators, B: Chemical 2003, 90, 68–75.
49. Xu, S.; Cai, X.; Tan, X.; Zhu, Y.; Lu, B. A fiber-optic biosensor using aptamer as receptors. Proceedings of SPIE - The International Society for Optical Engineering; 2001; 4414, pp. 35–37.
50. Watterson, J. H.; Piunno, P. A. E.; Wust, C. C.; Krull, U. J. Controlling the density of nucleic acid oligomers on fiber optic sensors for enhancement of selectivity and sensitivity. Sensors and Actuators, B: Chemical 2001, 74, 27–36.
51. Pilevar, S.; Davis, C. C.; Portugal, F. Tapered optical fiber sensor using near-infrared fluorophores to assay hybridization. Analytical Chemistry 1998, 70, 2031–2037.
52. Pilevar, S.; Davis, C. C.; Hodzic, V.; Portugal, F. Optical fiber hybridization assay fluorosensor. Proceedings of SPIE - The International Society for Optical Engineering; 1999; 3666, pp. 585–588.
53. Hanafi-Bagby, D.; Piunno, P. A. E.; Wust, C. C.; Krull, U. J. Concentration dependence of a thiazole orange derivative that is used to determine nucleic acid hybridization by an optical biosensor. Analytica Chimica Acta 2000, 411, 19–30.
54. Tatsu, Y.; Yamamura, S.; Yoshikawa, S. Fluorescent fibre-optic immunosensing system based on complement lysis of liposome

containing carboxyfluorescein. Biosensors and Bioelectronics 1992, 7(6), 741–745.
55. Jiang, L.; Margaritis, A.; Petersen, N. O. Development of a new optical fiber epifluorescence biosensor system. Proceedings of SPIE - The International Society for Optical Engineering; 1992; 1648, pp. 212–222.
56. Anderson, G. P.; King, K. D.; Cao, L. K.; Jacoby, M.; Ligler, F. S.; Ezzell, J. Quantifying serum antiplague antibody with a fiber-optic biosensor. Clinical and Diagnostic Laboratory Immunology 1998, 5, 609–612.
57. Balcer, H. I.; Kwon, H. J.; Kang, K. A. Assay procedure optimization of a rapid, reusable protein c immunosensor for physiological samples. Annals of Biomedical Engineering 2002, 30, 141–147.
58. Shriver-Lake, L. C.; Donner, B.; Edelstein, R.; Breslin, K.; Bhatia, S. K.; Ligler, F. S. Antibody immobilization using heterobifunctional crosslinkers. Biosensors and Bioelectronics 1997, 12, 1101–1106.
59. Clark, H. A.; Merritt, G.; Kopelman, R. Novel optical biosensors using a gold colloid monolayer substrate. Proceedings of SPIE - The International Society for Optical Engineering; 2000; 3922, pp. 138–146.
60. Thompson, R. B.; Jones, E. R. Enzyme-based fiber optic zinc biosensor. Analytical Chemistry 1993, 65, 730–733.
61. Thompson, R. B.; Ge, Z.; Patchan, M. W.; Fierke, C. A. Energy-transfer-based fiber optic metal-ion biosensor. Proceedings of SPIE - The International Society for Optical Engineering; 1995; 2388, pp. 138–147.
62. Chang, C. J.; Nolan, E. M.; Jaworski, J.; Okamoto, K. I.; Hayashi, Y.; Sheng, M.; Lippard, S. J. ZP8, a neuronal zinc sensor with improved dynamic range; Imaging zinc in hippocampal slices with two-photon microscopy. Inorganic Chemistry 2004, 43, 6774–6779.
63. Anderson, G. P.; Breslin, K. A.; Ligler, F. S. Assay development for a portable fiberoptic biosensor. ASAIO Journal 1996, 42, 942–946.
64. Chen, Z.; Kaplan, D. L.; Yang, K.; Kumar, J.; Marx, K. A.; Tripathy, S. K. Phycobiliproteins encapsulated in sol-gel glass. Journal of Sol-Gel Science and Technology 1996, 7, 99–108.

65. Flora, K.; Brennan, J. D. Comparison of formats for the development of fiber-optic biosensors utilizing sol-gel derived materials entrapping fluorescently-labelled protein. Analyst 1999, 124, 1455–1462.
66. Doong, R. A.; Tsai, H. C. Immobilization and characterization of sol-gel-encapsulated acetylcholinesterase fiber-optic biosensor. Analytica Chimica Acta 2001, 434, 239–246.
67. Anderson, G. P.; Jacoby, M. A.; Ligler, F. S.; King, K. D. Effectiveness of protein a for antibody immobilization for a fiber optic biosensor. Biosensors and Bioelectronics 1997, 12, 329–336.
68. Navas Diaz, A.; Ramos Peinado, M. C. Sol-Gel cholinesterase biosensor for organophosphorus pesticide fluorimetric analysis. Sensors and Actuators, B: Chemical 1997, 39, 426–431.
69. Choa, F. S.; Shih, M. H.; Toppozada, A. R.; Block, M.; Eldefrawi, M. E. A light emitting diode improves evanescent excitation of a fiber optic cocaine biosensor. Analytical Letters 1996, 29, 29–33.
70. Healey, B. G.; Li, L.; Walt, D. R. Multianalyte biosensors on optical imaging bundles. Biosensors and Bioelectronics 1997, 12, 521–529.
71. Xavier, M. P.; Garcia-Fresnadillo, D.; Moreno-Bondi, M. C.; Orellana, G. Oxygen sensing in nonaqueous media using porous glass with covalently bound luminescent Ru(II) complexes. Analytical Chemistry 1998, 70, 5184–5189.
72. Barker, S. L. R.; Clark, H. A.; Swallen, S. F.; Kopelman, R.; Tsang, A. W.; Swanson, J. A. Ratiometric and fluorescence-lifetime-based biosensors incorporating cytochrome c' and the detection of extra- and intracellular macrophage nitric oxide. Analytical Chemistry 1999, 71, 1767–1772.
73. Rabinovich, E. M.; Svimonishvili, T.; O'Brien, M. J.; Brueck, S. R. J.; Buranda, T.; Sklar, L. A.; Lopez, G. P. Development of a new detection technique for fluorescence lifetime-based chemical/biological sensor arrays monitoring: Dual closed-loop optoelectronic auto-oscillatory detection circuit. Proceedings of SPIE - The International Society for Optical Engineering; 2002; 4624, pp. 115–122.
74. Niu, C. G.; Li, Z. Z.; Zhang, X. B.; Lin, W. Q.; Shen, G. L.; Yu, R. Q. Covalently immobilized aminonaphthalimide as fluorescent carrier

for the preparation of optical sensors. Analytical and Bioanalytical Chemistry 2002, 372, 519–524.

75. Speckman, D. M.; Jennings, T. L.; La Lumondiere, S. D.; Moss, S. C. Development of quantum dot reporters for immunoassay applications. Materials Research Society Symposium -Proceedings; 2001; 676.
76. Speckman, D. M.; Jennings, T. L.; LaLumondiere, S. D.; Klimcak, C. M.; Moss, S. C.; Loper, G. L.; Beck, S. M. Biodetection using fluorescent quantum dots. Proceedings of SPIE - The International Society for Optical Engineering; 2002; 4745, pp. 136–143.
77. Walczak, I. M.; Love, W. F.; Cook, T. A.; Slovacek, R. E. The application of evanescent wave sensing to a high-sensitivity fluoroimmunoassay. Biosensors and Bioelectronics 1992, 7, 39–48.
78. Weber, A.; Schultz, J. S. Fiber-optic fluorimetry in biosensors: Comparison between evanescent wave generation and distal-face generation of fluorescent light. Biosensors and Bioelectronics 1992, 7, 193–197.
79. Anderson, G. P.; Golden, J. P.; Ligler, F. S. A fiber optic biosensor: Combination tapered fibers designed for improved signal acquisition. Biosensors and Bioelectronics 1993, 8, 249–256.
80. Golden, J. P.; Rabbany, S. Y.; Anderson, G. P. Ray-tracing determination of evanescent-wave penetration depth in tapered fiber optic probes. Proceedings of SPIE - The International Society for Optical Engineering; 1993; 1796, pp. 9–13.
81. Anderson, G. P.; Golden, J. P.; Cao, L. K.; Wijesuriya, D.; Shriver-Lake, L. C.; Ligler, F. S. Development of an evanescent wave fiber optic biosensor. IEEE Engineering in Medicine and Biology Magazine 1994, 13, 358–363.
82. Anderson, G. P.; Golden, J. P.; Ligler, F. S. Evanescent wave biosensor. Part I. Fluorescent signal acquisition from step-etched fiber optic probes. IEEE Transactions on Biomedical Engineering 1994, 41, 578–589.
83. Golden, J.P.; Anderson, G. P.; Rabbany, S.Y.; Ligler, F.S. An evanescent wave biosensor-Part II: Fluorescence signal acquisition from tapered fiber optic probes. IEEE Transactions on Biomedical Engineering 1994, 41, 585–591.

84. Feldman, S. F.; Uzgiris, E. E.; Murray Penney, C.; Gui, J. Y.; Shu, E. Y.; Stokes, E. B. Evanescent wave immunoprobe with high bivalent antibody activity. Biosensors and Bioelectronics 1995, 10, 423–434.
85. Bronk, K. S.; Michael, K. L.; Pantano, P.; Walt, D. R. Combined imaging and chemical sensing using a single optical imaging fiber. Analytical Chemistry 1995, 67, 2750–2757.
86. Hale, Z. M.; Payne, F. P.; Marks, R. S.; Lowe, C. R.; Levine, M. M. The single mode tapered optical fibre loop immunosensor. Biosensors and Bioelectronics 1996, 11, 137–148.
87. Golden, J. P.; Saaski, E. W.; Shriver-Lake, L. C.; Anderson, G. P.; Ligler, F. S. Portable multichannel fiber optic biosensor for field detection. Optical Engineering 1997, 36, 1008–1013.
88. Fielding, A. J.; Davis, C. C. Tapered single-mode optical fiber evanescent coupling. IEEE Photonics Technology Letters 2002, 14, 53–55.
89. Huijie, H.; Junhui, Z.; Yongkai, Z.; Ruifu, Y.; Bingqiang, R.; Zhaogu, C.; Longlong, D.; Dunwu, L. Design and application of fiber-optic evanescent wave biosensor. Proceedings of SPIE - The International Society for Optical Engineering; 2003; 5254, pp. 316–322.
90. Garden, S. R.; Doellgast, G. J.; Killham, K. S.; Strachan, N. J. C. A fluorescent coagulation assay for thrombin using a fibre optic evanescent wave sensor. Biosensors and Bioelectronics 2004, 19, 737–740.
91. Huang, H. J.; Zhai, J. H.; Zhao, Y. K.; Yang, R. F.; Ren, B. Q.; Cheng, Z. G.; Du, L. L.; Lu, D. W. Multi-probe fiber-optic evanescent wave biosensor and its characterization. Zhongguo Jiguang/Chinese Journal of Lasers 2004, 31, 718–722.
92. Liao, K. C.; Richmond, F. J.; Hogen-Esch, T.; Marcu, L.; Loeb, G. E. Sencil™ project: Development of a percutaneous optical biosensor. Annual International Conference of the IEEE Engineering in Medicine and Biology - Proceedings; 2004; 26 III, pp. 2082–2085.
93. Ahmad, M.; Hench, L.L. Effect of taper geometrices and launch angle on evanescent wave penetration depth in optical fibers. Biosensors and Bioelectronics 2005, 20, 1312–1319.

94. Deng, L. X.; Feng, Y. Structure study of fiber optical evanescent wave biosensor. Chinese Journal of Sensors and Actuators 2005, 18, 353–357.
95. Chao, K. F.; Zhang, Y. L.; Kong, X. G.; Zeng, Q. H.; Liu, R. L.; Wang, X.; Feng, L. Y.; Sun, Y. J. A study of interaction between fitc- labeled antibody and optical fiber surface. Proceedings of SPIE - The International Society for Optical Engineering; 2006; 6029.
96. Hong, B.; Kang, K. A. Biocompatible, nanogold-particle fluorescence enhancer for fluorophore mediated, optical immunosensor. Biosensors and Bioelectronics 2006, 21, 1333–1338.
97. Emiliyanov, G.; Jensen, J. B.; Bang, O.; Hoiby, P. E.; Pedersen, L. H.; Kjær, E. M.; Lindvold, L. Localized biosensing with topas microstructured polymer optical fiber. Optics Letters 2007, 32, 460–462.
98. Anderson, G. P.; Shriver-Lake, L. C.; Golden, J. P.; Ligler, F. S. Fiber-optic-based biosensor: signal enhancement in a production model. Proceedings of SPIE - The International Society for Optical Engineering; 1992; 1648, pp. 39–43.
99. Piunno, P. A. E.; Krull, U. J.; Hudson, R. H. E.; Damha, M. J.; Cohen, H. Fiber optic biosensor for fluorimetric detection of dna hybridization. Analytica Chimica Acta 1994, 288, 205–214.
100. Piunno, P. A. E.; Krull, U. J.; Hudson, R. H. E.; Damha, M. J.; Cohen, H. Fiber-Optic DNA sensor for fluorometric nucleic acid determination. Analytical Chemistry 1995, 67, 2635–2643.
101. Nguyen-Ngoc, H.; Tran-Minh, C. Fluorescent biosensor using whole cells in an inorganic translucent matrix. Analytica Chimica Acta 2007, 583, 161–165.
102. Wygladacz, K.; Bakker, E. Imaging fiber microarray fluorescent ion sensors based on bulk optode microspheres. Analytica Chimica Acta 2005, 532, 61–69.
103. Ayyagari, M. S.; Pande, R.; Kamtekar, S.; Gao, H.; Marx, K. A.; Kumar, J.; Tripathy, S. K.; Akkara, J. A.; Kaplan, D. L. Molecular assembly of proteins and conjugated polymers: Toward development of biosensors. Biotechnology and Bioengineering 1995, 45, 116–121.

104. Chen, Z.; Gao, H. H.; Kumar, J.; Marx, K. A.; Tripathy, S. K.; Kaplan, D. L. Molecular assembly of multilayer enzyme on fiber optic surface: Toward the development of chemiluminescence-based fiber optic biosensors. Proceedings of SPIE - The International Society for Optical Engineering; 1996; 2676, pp. 156–164.
105. Gao, H. H.; Chen, Z.; Kumar, J.; Marx, K. A.; Tripathy, S. K.; Kaplan, D. L. Multilayer enzyme assembly for the development of a novel fiber optic biosensor. Materials Research Society Symposium - Proceedings; 1996; 414, pp. 125–130.
106. Chen, Z.; Kaplan, D. L.; Gao, H.; Kumar, J.; Marx, K. A.; Tripathy, S. K. Molecular assembly of multilayer enzyme: toward the development of a chemiluminescence-based fiber optic biosensor. Materials Science and Engineering C 1996, 4, 155–159.
107. Marquette, C. A.; Blum, L. J. Luminol electrochemiluminescence-based fibre optic biosensors for flow injection analysis of glucose and lactate in natural samples. Analytica Chimica Acta 1999, 381, 1–10.
108. Marquette, C. A.; Leca, B. D.; Blum, L. J. Electrogenerated chemiluminescence of luminol for oxidase-based fibre-optic biosensors. Luminescence 2001, 16, 159–165.
109. Palaroan, W. S.; Bergantin, J., Jr.; Sevilla, F., III. Optical fiber chemiluminescence biosensor for antioxidants based on an immobilized luminol/hematin reagent phase. Analytical Letters 2000, 33, 1797–1810.
110. Ramos, M. C.; Torijas, M. C.; Navas Diaz, A. Enhanced chemiluminescence biosensor for the determination of phenolic compounds and hydrogen peroxide. Sensors and Actuators, B: Chemical 2001, 73, 71–75.
111. Li, Y. X.; Zhu, L. D.; Zhu, G. Y.; Zhao, C. Chemiluminescence optical fiber glucose biosensor based on co-immobilizing glucose oxidase and horseradish peroxidase in a sol-gel film. Chemical Research in Chinese Universities 2002, 18, 12–15.
112. Magrisso, M.; Etzion, O.; Pilch, G.; Novodvoretz, A.; Perez-Avraham, G.; Schlaeffer, F.; Marks, R. Fiber-optic biosensor to assess circulating phagocyte activity by chemiluminescence. Biosensors and Bioelectronics 2006, 21, 1210–1218.

113. Gautier, S. M.; Blum, L. J.; Coulet, P. R. Fibre-optic biosensor based on luminescence and immobilized enzymes: microdetermination of sorbitol, ethanol and oxaloacetate. Journal of bioluminescence and chemiluminescence 1990, 5, 57–63.
114. Gautier, S. M.; Blum, L. J.; Coulet, P. R. Multi-function fibre-optic sensor for the bioluminescent flow determinationof ATP or NADH. Analytica Chimica Acta 1990, 235, 243–253.
115. Coulet, P. R. Polymeric membranes and coupled enzymes in the design of biosensors. Journal of Membrane Science 1992, 68, 217–228.
116. Klee, B.; Duveneck, G. L.; Oroszlan, P.; Ehrat, M.; Widmer, H. M. Model system for the development of an optical biosensor based on lipid membranes and membrane-bound receptors. Sensors and Actuators, B: Chemical 1995, B29, 307–311.
117. Blum, L. J.; Gautier, S. M.; Berger, A.; Michel, P. E.; Coulet, P. R. Multicomponent organized bioactive layers for fiber-optic luminescent sensors. Sensors and Actuators, B: Chemical 1995, B29, 1–9.
118. Michel, P. E.; Gautier-Sauvigne, S. M.; Blum, L. J. A transient enzymatic inhibition as an efficient tool for the discriminating bioluminescent analysis of three adenylic nucleotides with a fiberoptic sensor based on a compartmentalized tri-enzymatic sensing layer. Analytica Chimica Acta 1998, 360, 89–99.
119. Choi, S. H.; Gu, M. B. A portable toxicity biosensor using freeze-dried recombinant bioluminescent bacteria. Biosensors and Bioelectronics 2002, 17, 433–440.
120. Polyak, B.; Geresh, S.; Marks, R. S. Synthesis and characterization of a biotin-alginate conjugate and its application in a biosensor construction. Biomacromolecules 2004, 5, 389–396.
121. Gruzina, T. G.; Vember, V. V.; Ul'berg, Z. R.; Starodub, N. F. Luminescent test based on photobacterium phosphorum b7071 for ionic and colloid gold determination in water. Khimiya i Tekhnologiya Vody 2005, 27, 200–208.
122. Eltoukhy, H.; Salama, K.; Gamal, A. E. A 0.18-Mm CMOS bioluminescence detection Lab-on-Chip. IEEE Journal of Solid-State Circuits 2006, 41, 651–661.

123. Papkovsky, D. B.; Olah, J.; Troyanovsky, I. V.; Sadovsky, N. A.; Rumyantseva, V. D.; Mironov, A. F.; Yaropolov, A. I.; Savitsky, A. P. Phosphorescent polymer films for optical oxygen sensors. Biosensors and Bioelectronics 1992, 7, 199–206.
124. Ovchinnikov, A. N.; Ogurtsov, V. I.; Trettnak, W.; Papkovsky, D. B. Enzymatic flow-injection analysis of metabolites using new type of oxygen sensor membranes and phosphorescence phase measurements. Analytical Letters 1999, 32, 701–716.
125. Akkin, T.; Dave, D. P.; Milner, T. E.; Rylander, H. G., III. Interferometric fiber-based optical biosensor to measure ultra-small changes in refractive index. Proceedings of SPIE - The International Society for Optical Engineering; 2002; 4616, pp. 9–13.
126. Tubb, A. J. C.; Payne, F. P.; Millington, R. B.; Lowe, C. R. Single-mode optical fibre surface plasma wave chemical sensor. Sensors and Actuators, B: Chemical 1997, 41, 71–79.
127. Pennington, C.; Jones, M.; Evans, M.; VanTassell, R.; Averett, J. Fiber optic based biosensors utilizing long period grating (LPG) technology. Proceedings of SPIE - The International Society for Optical Engineering; 2001; 4255, pp. 53–62.
128. Cheng, S. F.; Chau, L. K. Colloidal gold-modified optical fiber for chemical and biochemical sensing. Analytical Chemistry 2003, 75, 16–21.
129. Tang, J. L.; Cheng, S. F.; Hsu, W. T.; Chiang, T. Y.; Chau, L. K. Fiber-optic biochemical sensing with a colloidal gold-modified long period fiber grating. Sensors and Actuators, B: Chemical 2006, 119, 105–109.
130. Piliarik, M.; Homola, J.; Maníková, Z.; Čtyroký, J. Surface plasmon resonance sensor based on a single-mode polarization-maintaining optical fiber. Sensors and Actuators, B: Chemical 2003, 90, 236–242.
131. Mitsui, K.; Handa, Y.; Kajikawa, K. Optical fiber affinity biosensor based on localized surface plasmon resonance. Applied Physics Letters 2004, 85, 4231–4233.
132. Guo, K. M.; Cheng, S. F.; Chau, L. K.; Tang, J. L. Optical biosensor based on colloidal gold-modified long-period fiber grating. Second Asian and Pacific Rim Symposium on Biophotonics -Proceedings, APBP 2004; 2004; pp. 207–208.

133. Chang, Y. J.; Chen, Y. C.; Kuo, H. L.; Wei, P. K. Nanofiber optic sensor based on the excitation of surface plasmon wave near fiber tip. Journal of Biomedical Optics 2006, 11, art. n°. 014032.
134. Wei, P. K.; Chang, Y. J.; Lee, C. W. Sensitive plasmonic biosensor using gold nanoparticles on a nano fiber tip. Progress in Biomedical Optics and Imaging - Proceedings of SPIE; 2006; 6099.
135. Lin, H. Y.; Tsai, W. H.; Tsao, Y. C.; Sheu, B. C. Side-polished multimode fiber biosensor based on surface plasmon resonance with halogen light. Applied Optics 2007, 46, 800–806.
136. Kurihara, K.; Ohkawa, H.; Iwasaki, Y.; Niwa, O.; Tobita, T.; Suzuki, K. Fiber-optic conical microsensors for surface plasmon resonance using chemically etched single-mode fiber. Analytica Chimica Acta 2004, 523, 165–170.
137. Dagenais, M.; Chryssis, A. N.; Yi, H.; Lee, S. M.; Saini, S. S.; Bentley, W. E. Optical biosensors based on etched fiber bragg gratings. Progress in Biomedical Optics and Imaging -Proceedings of SPIE; 2005; 5729, pp. 214–224.
138. Chryssis, A. N.; Lee, S. M.; Lee, S. B.; Saini, S. S.; Dagenais, M. High sensitivity evanescent field fiber bragg grating sensor. IEEE Photonics Technology Letters 2005, 17, 1253–1255.
139. Sang, X. Z.; Yu, C. X.; Yan, B. B.; Mayteevarunyoo, T.; Lu, N. G.; Zhang, Q. Chemical sensor based on a fiber bragg grating. Guangxue Jingmi Gongcheng/Optics and Precision Engineering 2006, 14, 771–774.
140. Chen, X.; Zhou, K.; Zhang, L.; Bennion, I. Dual-peak long-period fiber gratings with enhanced refractive index sensitivity by finely tailored mode dispersion that uses the light cladding etching technique. Applied Optics 2007, 46, 451–455.
141. Jung, Y.; Kim, S.; Lee, D. S.; Oh, K. Compact three segmented multimode fiber modal interferometer for high sensitivity refractive-index measurement. Proceedings of SPIE - The International Society for Optical Engineering; 2005; 5855 PART I, pp. 150–153.
142. Chandrasekaran, A.; Packirisamy, M. MOEMS based integrated microfluidic fiber-optic waveguides for biophotonic applications.

Progress in Biomedical Optics and Imaging -Proceedings of SPIE; 2005; 5969.

143. Pepper, D. M. Use of phase-conjugate interferometry for trace-compound detection. Proceedings of SPIE - The International Society for Optical Engineering; 1993; 1824, pp. 79–96.
144. Wavering, T.; Meller, S.; Evans, M.; Pennington, C.; Jones, M.; Van Tassell, R.; Murphy, K.; Velander, W.; Valdes, E. Interferometric optical fiber microcantilever beam biosensor. Proceedings of SPIE - The International Society for Optical Engineering; 2000; 4200, pp. 10–16.
145. Chiu, M. H.; Wang, S. F.; Chang, R. S. D-type fiber biosensor based on surface-plasmon resonance technology and heterodyne interferometry. Optics Letters 2005, 30, 233–235.
146. Zang, Y.; Shibru, H.; Cooper, K.L.; Wang, A. Miniature fiber-optic multicavity Fabry-Perot interferometric biosensor. Optics Letters 2005, 30, 1021–1023.
147. Yang, Y.; Zhang, B.; Wang, Y. Study of key technology of optical fiber protein chips. Proceedings of SPIE - The International Society for Optical Engineering; 2006; 6150 II.
148. Rindorf, L.; Høiby, P. E.; Jensen, J. B.; Pedersen, L. H.; Bang, O.; Geschke, O. Towards biochips using microstructured optical fiber sensors. Analytical and Bioanalytical Chemistry 2006, 385, 1370–1375.
149. Sadana, A.; Alari, J. P.; Vo-dinh, T. A β-cyclodextrin based fiber-optic chemical sensor: a fractal analysis. Talanta 1995, 42, 1567–1574.
150. Sadana, A.; Sutaria, M. Influence of diffusion to fractal surfaces on the binding kinetics for antibody-antigen, analyte-receptor, and analyte-receptorless (protein) systems. Biophysical Chemistry 1997, 65, 29–44.
151. Doke, A.; Mathur, S.; Sadana, A. Fractal binding and dissociation kinetics of heart-related compounds on biosensor surfaces. Journal of Receptors and Signal Transduction 2006, 26, 337–357.
152. Noto, M.; Khoshsima, M.; Keng, D.; Teraoka, I.; Kolchenko, V.; Arnold, S. Molecular weight dependence of a whispering gallery mode biosensor. Applied Physics Letters 2005, 87, 1–3.

153. Taylor, L. C.; Walt, D. R. Application of high-density optical microwell arrays in a live-cell biosensing system. Analytical Biochemistry 2000, 278, 132–142.
154. Walt, D. R. Imaging optical sensor arrays. Current Opinion in Chemical Biology 2002, 6, 689–695.
155. Deverkadra, R. R.; McShane, M. J. Fabrication of a biosensor with embedded waveguides for sensing in microchannels. Annual International Conference of the IEEE Engineering in Medicine and Biology - Proceedings; 2002; 2, pp. 1720–1721.
156. Konry, T.; Novoa, A.; Cosnier, S.; Marks, R. S. Development of an "electroptode" immunosensor: indium tin oxide-coated optical fiber tips conjugated with an electropolymerized thin film with conjugated cholera toxin B subunit. Analytical Chemistry 2003, 75, 2633–2639.
157. Konry, T.; Marks, R. S. Physico-chemical studies of indium tin oxide-coated fiber optic biosensors. Thin Solid Films 2005, 492, 313–321.
158. Brogan, K. L.; Walt, D. R. Optical fiber-based sensors: Application to chemical biology. Current Opinion in Chemical Biology 2005, 9, 494–500.
159. Vo-Dinh, T.; Griffin, G. D.; Alarie, J. P.; Cullum, B.; Sumpter, B.; Noid, D. Development of nanosensors and bioprobes. Journal of Nanoparticle Research 2000, 2, 17–27.
160. Vo-Dinh, T. Nanobiosensors: Probing the sanctuary of individual living cells. Journal of Cellular Biochemistry 2002, 154–161.
161. Wadkins, R. M.; Golden, J. P.; Ligler, F. S. Calibration of biosensor response using simultaneous evanescent wave excitation of cyanine-labeled capture antibodies and antigens. Analytical Biochemistry 1995, 232, 73–78.
162. Anderson, G. P.; Nerurkar, N. L. Improved Fluoroimmunoassays using the dye Alexa Fluor 647 with the RAPTOR, a fiber optic biosensor. Journal of Immunological Methods 2002, 271, 17–24.
163. Huang, H.; Zhai, J.; Ren, B.; Yang, R.; Zhao, Y.; Cheng, Z.; Du, L.; Lu, D. Fiber-optic evanescent wave biosensor and its application. Guangxue Xuebao/Acta Optica Sinica 2003, 23, 451–454.

164. Anderson, G. P.; King, K. D.; Cuttino, D. S.; Whelan, J. P.; Ligler, F. S.; MacKrell, J. F.; Bovais, C. S.; Indyke, D. K.; Foch, R. J. Biological agent detection with the use of an airborne biosensor. Field Analytical Chemistry and Technology 1999, 3, 307–314.
165. King, K. D.; Vanniere, J. M.; Leblanc, J. L.; Bullock, K. E.; Anderson, G. P. Automated fiber optic biosensor for multiplexed immunoassays. Environmental Science and Technology 2000, 34, 2845–2850.
166. Anderson, G. P.; Rowe-Taitt, C. A. Water quality monitoring using an automated portable fiber optic biosensor: RAPTOR. Proceedings of SPIE - The International Society for Optical Engineering; 2001; 4206, pp. 58–63.
167. Tu, S. I.; Geng, T.; Uknalis, J.; Bhunia, A. fiber optic biosensor employing Alexa-Fluor conjugated antibodies for detection of Escherichia Coli O157:H7 and Shiga-like toxins. Proceedings of SPIE - The International Society for Optical Engineering; 2006; 6381.
168. Rijal, K.; Leung, A.; Shankar, P. M.; Mutharasan, R. Detection of pathogen Escherichia Coli O157:H7 at 70 cells/mL using antibody-immobilized biconical tapered fiber sensors. Biosensors and Bioelectronics 2005, 21, 871–880.
169. DeMarco, D. R.; Lim, D. V. Detection of Escherichia Coli O157:H7 in 10- and 25-Gram ground beef samples with an evanescent-wave biosensor with silica and polystyrene waveguides. Journal of Food Protection 2002, 65, 596–602.
170. Chanda, R.; Irudayaraj, J.; Pantano, C. G. Characterization of thin films for optical sensors of food-borne pathogens. Proceedings of SPIE - The International Society for Optical Engineering; 2004; 5591, pp. 231–238.
171. Chang, A. C.; Gillespie, J. B.; Tabacco, M. B. Enhanced detection of live bacteria using a dendrimer thin film in an optical biosensor. Analytical Chemistry 2001, 73, 467–470.
172. Chang, A. C.; Tabacco, M. B. Real-time detection of bacterial aerosols using fiber optic-based biosensors. Proceedings of SPIE - The International Society for Optical Engineering; 2001; 4206, pp. 40–47.

173. Chuang, H.; Macuch, P.; Tabacco, M. B. Optical sensors for detection of bacteria. 1. General concepts and initial development. Analytical Chemistry 2001, 73, 462–466.
174. Chang, Y. H.; Chang, T. C.; Kao, E. F.; Chou, C. Detection of protein A produced by Staphylococcus Aureus with a fiber-optic-based biosensor. Bioscience, Biotechnology and Biochemistry 1996, 60, 1571–1574.
175. Ferreira, A. P.; Werneck, M. M.; Ribeiro, R. M. Aerobiological pathogen detection by evanescent wave fibre optic sensor. Biotechnology Techniques 1999, 13, 447–452.
176. Ferreira, A. P.; Werneck, M. M.; Ribeiro, R. M.; Lins, U. G. C. Pathogen detection using evanescent wave fiber optic biosensor. Proceedings of SPIE - The International Society for Optical Engineering; 1999; 3749, pp. 395–396.
177. Bhunia, A. K.; Geng, T.; Lathrop, A.; Valadez, A.; Morgan, M. T. Optical immunosensors for detection of listeria monocytogenes and salmonella enteritidis from food. Proceedings of SPIE -The International Society for Optical Engineering; 2004; 5271, pp. 1–6.
178. Ko, S.; Grant, S. A. Development of a novel FRET immunosensor for detection of listeria. Proceedings of IEEE Sensors; 2003; 2, pp. 288–292.
179. Ko, S.; Grant, S. A. A novel FRET-based optical fiber biosensor for rapid detection of Salmonella typhimurium. Biosensors and Bioelectronics 2006, 21, 1283–1290.
180. Zhou, C.; Pivarnik, P.; Auger, S.; Rand, A.; Letcher, S. A compact fiber-optic immunosensor using tapered fibers and acoustic enhancement. Proceedings of SPIE - The International Society for Optical Engineering; 1997; 2976, pp. 40–50.
181. Biran, I.; Rissin, D. M.; Walt, D. R. Optical imaging fiber based recombinant bacterial biosensor. Proceedings of SPIE - The International Society for Optical Engineering; 2001; 4576, pp. 68–74.
182. Cao, L. K.; Anderson, G. P.; Ligler, F. S.; Ezzell, J. Detection of Yersinia pestis fraction 1 antigen with a fiber optic biosensor. Journal of Clinical Microbiology 1995, 33, 336–341.

183. Zhai, J.; Huang, H.; Yang, R.; Ren, B.; Zhao, Y. A fiber-optic evanescent wave biosensor for the specific detection of nucleic acids. Fenxi Huaxue 2003, 31, 34–37.
184. Zhang, G.; Zhou, Y.; Yuan, J.; Ren, S. A Chemiluminescence fiber-optic biosensor for detection of DNA hybridization. Analytical Letters 1999, 32, 2725–2736.
185. Watterson, J. H.; Piunno, P. A. E.; Krull, U. J. Towards the optimization of an optical DNA sensor: Control of selectivity coefficients and relative surface affinities. Analytica Chimica Acta 2002, 457, 29–38.
186. Wang, X.; Krull, U. J. Synthesis and fluorescence studies of thiazole orange tethered onto oligonucleotide: development of a self-contained DNA biosensor on a fiber optic surface. Bioorganic and Medicinal Chemistry Letters 2005, 15, 1725–1729.
187. Wang, X.; Krull, U. J. A self-contained fluorescent fiber optic DNA biosensor. Journal of Materials Chemistry 2005, 15, 2801–2809.
188. Uddin, A. H.; Piunno, P. A. E.; Hudson, R. H. E.; Damha, M. J.; Krull, U. J. A fiber optic biosensor for fluorimetric detection of triple-helical DNA. Nucleic Acids Research 1997, 25, 4139–4146.
189. Rogers, K. R.; Apostol, A.; Madsen, S. J.; Spencer, C. W. Fiber optic biosensor for detection of DNA damage. Analytica Chimica Acta 2001, 444, 51–60.
190. Lu, H.; Zhao, Y.; Ma, J.; Lu, Z. Photon counting optical fiber biosensor for the characterization of nucleic acid hybridization. Proceedings of SPIE - The International Society for Optical Engineering; 1999; 3863, pp. 138–142.
191. Liu, X.; Tan, W. A fiber-optic evanescent wave DNA biosensor based on novel molecular beacons. Analytical Chemistry 1999, 71, 5054–5059.
192. Liu, X.; Farmerie, W.; Schuster, S.; Tan, W. Molecular beacons for DNA biosensors with micrometer to submicrometer dimensions. Analytical Biochemistry 2000, 283, 56–63.
193. Li, J.; Tan, W.; Wang, K.; Xiao, D.; Yang, X.; He, X.; Tang, Z. Ultrasensitive optical DNA biosensor based on surface immobilization of molecular beacon by a bridge structure. Analytical Sciences 2001, 17, 1149–1153.

194. Jiang, G.; Chen, R.; Yan, H.; Ouyang, Q. A new method of preparing fiber-optic DNA biosensor and its array for gene detection. Science in China, Series C: Life Sciences 2001, 44, 33–39.
195. Alarie, J. P.; Vo-Dinh, T. Antibody-based submicron biosensor for benzo[arsqb;pyrene DNA adduct. Polycyclic Aromatic Compounds 1996, 8, 45–52.
196. Cullum, B. M.; Griffin, G. D.; Miller, G. H.; Vo-Dinh, T. Intracellular measurements in mammary carcinoma cells using fiber-optic nanosensors. Analytical Biochemistry 2000, 277, 25–32.
197. Schubert, F. Fiber-optic enzyme sensor for the determination of adenosine diphosphate using internal analyte recycling. Sensors and Actuators, B: Chemical 1993, B11, 531–535.
198. Michel, P. E.; Gautier-Sauvigne, S. M.; Blum, L. J. Luciferin incorporation in the structure of acrylic microspheres with subsequent confinement in a polymeric film: A new method to develop a controlled release-based biosensor for ATP. ADP and AMP and Talanta 1998, 47, 169–181.
199. Chen, X.; Zhou, K.; Zhang, L.; Bennion, I. High sensitivity biosensor based on dual-peak LPG sensitised by light cladding etching. Proceedings of SPIE - The International Society for Optical Engineering; 2005; 5855 PART I, pp. 383–386.
200. Spiridonova, V. A.; Kopylov, A. M. DNA aptamers as radically new recognition elements for biosensors. Biochemistry 2002, 67, 706–709.
201. Schmidt, P. M.; Lehmann, C.; Matthes, E.; Bier, F. F. Detection of activity of telomerase in tumor cells using fiber optical biosensors. Biosensors and Bioelectronics 2002, 17, 1081–1087.
202. Yu, C. J.; Chou, C.; Hsu, H. Y.; Chan, T. S.; Lee, Z. Y.; Wu, H. T. Fiber optic biosensor of monitoring protein binding kinetics. Progress in Biomedical Optics and Imaging - Proceedings of SPIE; 2005; 5691, pp. 200–208.
203. Yu, S. H.; Chen, C. H.; Chiang, H. K. Fiber optic evanescent wave biosensor for biomolecular detection. Second Asian and Pacific Rim Symposium on Biophotonics - Proceedings, APBP 2004; 2004; pp. 165–166.

204. Henning, M. Fiber optic couplers for protein detection. Sensors and Materials 2002, 14, 339–345.
205. Poirier, M. A.; Lopes, T.; Singh, B. Use of an optical-fiber-based biosensor to study the interaction of blood proteins with solid surfaces. Applied Spectroscopy 1994, 48, 867–870.
206. Kang, K. A.; Anis, N. A.; Eldefrawi, M. E.; Drohan, W.; Bruley, D. F. Reusable, Real-time, immuno-optical protein C biosensor. Advances in Experimental Medicine and Biology 1997, 411, 437–444.
207. Spiker, J. O.; Kang, K. A.; Drohan, W. N.; Bruley, D. F. Preliminary study of biosensor optimization for the detection of protein C. Advances in Experimental Medicine and Biology 1998, 454, 681–688.
208. Spiker, J. O.; Kang, K. A. Preliminary study of real-time fiber optic based protein C biosensor. Biotechnology and Bioengineering 1999, 66, 158–163.
209. Preejith, P. V.; Lim, C. S.; Kishen, A.; John, M. S.; Asundi, A. Total protein measurement using a fiber-optic evanescent wave-based biosensor. Biotechnology Letters 2003, 25, 105–110.
210. Starodub, V. M.; Fedorenko, L. L.; Starodub, N. F. Optical immune sensors for the monitoring protein substances in the air. Sensors and Actuators, B: Chemical 2000, 68, 40–47.
211. Zhen, G.; Zurcher, S.; Falconnet, D.; Xu, F.; Kuennemann, E.; Textor, M. NTA-functionalized poly(l-lysine)-g-poly(ethylene glycol): A polymeric interface for binding and studying 6×His-tagged proteins. Annual International Conference of the IEEE Engineering in Medicine and Biology - Proceedings; 2005; 7, pp. 1036–1038.
212. Rickus, J. L.; Tobin, A. J.; Zink, J. I.; Dunn, B. Photochemical enzyme co-factor regeneration: Towards continuous glutamate monitoring with a sol-gel optical biosensor. Materials Research Society Symposium - Proceedings; 2002; 723, pp. 155–160.
213. Pierce, M. E.; Grant, S. A. Development of a FRET based fiber-optic biosensor for early detection of myocardial infarction. Annual International Conference of the IEEE Engineering in Medicine and Biology - Proceedings; 2004; 26 III, pp. 2098–2101.
214. Grant, S. A.; Pierce, M. E.; Lichlyter, D. J.; Grant, D. A. Effects of immobilization on a FRET immunosensor for the detection of

myocardial infarction. Analytical and Bioanalytical Chemistry 2005, 381, 1012–1018.
215. Grant, S. A.; Heits, B.; Kleiboeker, S. Development of an optical biosensor utilizing gold nanoparticles to detect porcine reproductive and respiratory syndrome virus. Sensor Letters 2006, 4, 246–252.
216. Petrosova, A.; Konry, T.; Cosnier, S.; Trakht, I.; Lutwama, J.; Rwaguma, E.; Chepurnov, A.; Muhlberger, E.; Lobel, L.; Marks, R. S. Development of a highly sensitive, field operable biosensor for serological studies of Ebola virus in central africa. Sensors and Actuators, B: Chemical 2007, 122, 578–586.
217. Konry, T.; Novoa, A.; Shemer-Avni, Y.; Hanuka, N.; Cosnier, S.; Lepellec, A.; Marks, R. S. Optical fiber immunosensor based on a poly(pyrrole-benzophenone) film for the detection of antibodies to viral antigen. Analytical Chemistry 2005, 77, 1771–1779.
218. Sansubrino, A.; Mascini, M. Development of an optical fibre sensor for ammonia, urea, urease and IgG. Biosensors and Bioelectronics 1994, 9, 207–216.
219. Li, X.; Fortuney, A.; Guilbault, G. G.; Suleiman, A. A. Determination of bilirubin by fiberoptic biosensor. Analytical Letters 1996, 29, 171–180.
220. Li, X.; Rosenzweig, Z. A fiber optic sensor for rapid analysis of bilirubin in serum. Analytica Chimica Acta 1997, 353, 263–273.
221. Vidal, M. M.; Delgadillo, I.; Gil, M. H.; Alonso-Chamarro, J. Study of an enzyme coupled system for the development of fibre optical bilirubin sensors. Biosensors and Bioelectronics 1996, 11, 347–354.
222. Pallotta, O. J.; Saccone, G. T. P.; Woolford, R. E. Bio-Sensor system discriminating between the biliary and pancreatic ductal systems. Medical and Biological Engineering and Computing 2006, 44, 250–255.
223. Bluestein, B. I.; Walczak, I. M.; Chen, S. Y. Fiber optic evanescent wave immunosensors for medical diagnostics. Trends in Biotechnology 1990, 8, 161–168.
224. Yang, X.; Xie, J. W.; Shen, G. L.; Yu, R. Q. An optical-fiber sensor for colchicine using photo-polymerized N-vinylcarbazole. Microchimica Acta 2003, 142, 225–230.

225. Marazuela, M. D.; Cuesta, B.; Moreno-Bondi, M. C.; Quejido, A. Free cholesterol fiber-optic biosensor for serum samples with simplex optimization. Biosensors and Bioelectronics 1997, 12, 233–240.
226. Yan, Y. H.; Xu, Y. H.; Li, S. P. Development of fiber optical biosensor of cholesterol based on fluorescence quenching. Chinese Journal of Biomedical Engineering 2004, 23, 44–48.
227. Marazuela, M. D.; Moreno-Bondi, M. C. Determination of choline-containing phospholipids in serum with a fiber-optic biosensor. Analytica Chimica Acta 1998, 374, 19–29.
228. Tsafack, V. C.; Marquette, C. A.; Leca, B.; Blum, L. J. An electrochemiluminescence-based fibre optic biosensor for choline flow injection analysis. Analyst 2000, 125, 151–155.
229. Takai, N.; Sakuma, I.; Fukui, Y.; Kaneko, A.; Fujie, T.; Taguchi, K.; Nagaoka, S. Studies of the development of optical fiber sensors for biochemical analysis. Artificial Organs 1991, 15, 86–89.
230. Schaffar, B. P. H.; Wolfbeis, O. S. A fast responding fibre optic glucose biosensor based on an oxygen optrode. Biosensors and Bioelectronics 1990, 5, 137–148.
231. Moreno-Bondi, M. C.; Wolfbeis, O. S.; Leiner, M. J. P.; Schaffar, B. P. H. Oxygen optrode for use in a fiber-optic glucose biosensor. Analytical Chemistry 1990, 62, 2377–2380.
232. Komives, C.; Schultz, J. S. Improvement of optical fiber based biosensors. Proceedings of the Annual Conference on Engineering in Medicine and Biology; 1990; pp. 478–479.
233. Del Villar, I.; Matias, I. R.; Arregui, F. J.; Corres, J. M. Fiber optic glucose biosensor. Optical Engineering 2006, 45.
234. Endo, H.; Yonemori, Y.; Musiya, K.; Maita, M.; Shibuya, T.; Ren, H.; Hayashi, T.; Mitsubayashi, K. A needle-type optical enzyme sensor system for determining glucose levels in fish blood. Analytica Chimica Acta 2006, 573-574, 117–124.
235. Jiang, D. S.; Liu, E.; Huang, J. Study of the complex enzyme membrane for a fiber optic glucose sensor. Key Engineering Materials 2003, 249, 425–428.
236. Kudo, K. Optical fiber bio-sensor using adsorption LB films. IEICE Transactions on Electronics 2004, E87-C, 185–187.

237. Lepore, M.; Portaccio, M.; Tommasi, E. D.; Luca, P. D.; Bencivenga, U.; Maiuri, P.; Mita, D. G. Glucose concentration determination by means of fluorescence emission spectra of soluble and insoluble glucose oxidase: Some useful indications for optical fibre-based sensors. Journal of Molecular Catalysis B: Enzymatic 2004, 31, 151–158.
238. Pasic, A.; Koehler, H.; Klimant, I.; Schaupp, L. Miniaturized fiber-optic hybrid sensor for continuous glucose monitoring in subcutaneous tissue. Sensors and Actuators, B: Chemical 2007, 122, 60–68.
239. Rosenzweig, Z.; Kopelman, R. Analytical Properties of miniaturized oxygen and glucose fiber optic sensors. Sensors and Actuators, B: Chemical 1996, 36, 475–483.
240. Rosenzweig, Z.; Kopelman, R. Analytical properties and sensor size effects of a micrometer-sized optical fiber glucose biosensor. Analytical Chemistry 1996, 68, 1408–1413.
241. Wolfbeis, O. S.; Oehme, I.; Papkovskaya, N.; Klimant, I. Sol-gel based glucose biosensors employing optical oxygen transducers, and a method for compensating for variable oxygen background. Biosensors and Bioelectronics 2000, 15, 69–76.
242. Zhu, L.; Li, Y.; Tian, F.; Xu, B.; Zhu, G. Electrochemiluminescent determination of glucose with a sol-gel derived ceramic-carbon composite electrode as a renewable optical fiber biosensor. Sensors and Actuators, B: Chemical 2002, 84, 265–270.
243. Young, J. S.; Scully, P. J.; Kvasnik, F.; Rose, K.; Kuncova, G.; Podrazky, O.; Matejec, V.; Mrazek, J. Optical fibre biosensors for oxygen and glucose monitoring. Proceedings of SPIE -The International Society for Optical Engineering; 2005; 5855 PART I, pp. 431–434.
244. Schubert, F.; Rinneberg, H. H.; Wang, F. Fiber optic biosensors for hydrogen peroxide and l-lactate. Proceedings of SPIE - The International Society for Optical Engineering; 1995; 2331, pp. 18–25.
245. McCormack, T.; O'Keeffe, G.; MacCraith, B. D.; O'Kennedy, R. Optical immunosensing of lactate dehydrogenase (LDH). Sensors and Actuators, B: Chemical 1997, 41, 89–96.

246. Grant, S. A.; Glass, R. S. Sol-gel-based biosensor for use in stroke treatment. IEEE Transactions on Biomedical Engineering 1999, 46, 1207–1211.
247. Xie, X.; Suleiman, A. A.; Guilbault, G. G. A fluorescence-based fiber optic biosensor for the flow-injection analysis of penicillin. Biotechnology and Bioengineering 1992, 39, 1147–1150.
248. Healey, B. G.; Walt, D. R. Development of a penicillin biosensor using a single optical imaging fiber. Proceedings of SPIE - The International Society for Optical Engineering; 1995; 2388, pp. 568–573.
249. Healey, B. G.; Walt, D. R. Improved fiber-optic chemical sensor for penicillin. Analytical Chemistry 1995, 67, 4471–4476.
250. Zhang, W.; Chang, H.; Rechnitz, G. A. Dual-enzyme fiber optic biosensor for pyruvate. Analytica Chimica Acta 1997, 350, 59–65.
251. Lee, S. J.; Scheper, T.; Buckmann, A. F. Application of a flow injection fibre optic biosensor for the analysis of different amino acids. Biosensors and Bioelectronics 1994, 9, 29–32.
252. Lu, W. X.; Chen, J. Continuous monitoring of adriamycin in vivo using fiber optic-based fluorescence chemical sensor. Analytical Chemistry 2003, 75, 1458–1462.
253. Merlo, S.; Yager, P.; Burgess, L. W. Optical method for detecting anesthetics and other lipid-soluble compounds. Sensors and Actuators, A: Physical 1990, 23, 1150–1154.
254. Nath, N.; Eldefrawi, M.; Wright, J.; Darwin, D.; Huestis, M. A rapid reusable fiber optic biosensor for detecting cocaine metabolites in urine. Journal of Analytical Toxicology 1999, 23, 460–467.
255. Toppozada, A. R.; Wright, J.; Eldefrawi, A. T.; Eldefrawi, M. E.; Johnson, E. L.; Emche, S. D.; Helling, C. S. Evaluation of a fiber optic immunosensor for quantitating cocaine in coca leaf extracts. Biosensors and Bioelectronics 1997, 12, 113–124.
256. Lundgren, J. S.; Bright, F. V. Biosensor for the nonspecific determination of ionic surfactants. Analytical Chemistry 1996, 68, 3377–3381.
257. Shriver-Lake, L. C.; Ogert, R. A.; Ligler, F. S. Fiber-optic evanescent-wave immunosensor for large molecules. Sensors and Actuators, B: Chemical 1993, B11, 239–243.

258. Slavik, R.; Brynda, E.; Homola, J.; Ctyroky, J. Miniature fiber optic surface plasmon resonance biosensor. Proceedings of SPIE - The International Society for Optical Engineering; 1999; 3570, pp. 184–191.
259. Slavík, R.; Homola, J.; Brynda, E. A miniature fiber optic surface plasmon resonance sensor for fast detection of staphylococcal enterotoxin B. Biosensors and Bioelectronics 2002, 17, 591–595.
260. Thompson, V. S.; Maragos, C. M. Fiber-optic immunosensor for the detection of fumonisin B1. Journal of Agricultural and Food Chemistry 1996, 44, 1041–1046.
261. Maragos, C. M.; Thompson, V. S. Fiber-optic immunosensor for mycotoxins. Natural Toxins 1999, 7, 371–376.
262. Ogert, R. A.; Brown, J. E.; Singh, B. R.; Shriver-Lake, L. C.; Ligler, F. S. Detection of Clostridium botulinum toxin A using a fiber optic-based biosensor. Analytical Biochemistry 1992, 205, 306–312.
263. Polyak, B.; Bassls, E.; Novodvorets, A.; Belkin, S.; Marks, R. S. Optical fiber bioluminescent whole-cell microbial biosensors to genotoxicants. Water Science and Technology 2000, 42, 305–311.
264. Narang, U.; Anderson, G. P.; Ligler, F. S.; Burans, J. Fiber optic-based biosensor for ricin. Biosensors and Bioelectronics 1997, 12, 937–945.
265. Mulchandani, A.; Kaneva, I.; Chen, W. Biosensor for direct determination of organophosphate nerve agents using recombinant Escherichia coli with surface-expressed organophosphorus hydrolase. 2. fiber-optic microbial biosensor. Analytical Chemistry 1998, 70, 5042–5046.
266. Mulchandani, A.; Pan, S.; Chen, W. Fiber-optic enzyme biosensor for direct determination of organophosphate nerve agents. Biotechnology Progress 1999, 15, 130–134.
267. Zhao, C. Q.; Anis, N. A.; Rogers, K. R.; Kline, R. H., Jr.; Wright, J.; Eldefrawi, A. T.; Eldefrawi, M. E. Fiber optic immunosensor for polychlorinated biphenyls. Journal of Agricultural and Food Chemistry 1995, 43, 2308–2315.
268. Wong, R. B.; Anis, N.; Eldefrawi, M. E. Reusable fiber-optic-based immunosensor for rapid detection of imazethapyr herbicide. Analytica Chimica Acta 1993, 279, 141–147.

269. Naessens, M.; Leclerc, J. C.; Tran-Minh, C. Fiber optic biosensor using Chlorella vulgaris for determination of toxic compounds. Ecotoxicology and Environmental Safety 2000, 46, 181–185.
270. Andreu, Y.; Baldini, F.; Domenici, C.; Giannetti, A.; Masci, D.; Mencaglia, A. Optical fibre sensor for photosynthetic herbicides detection by time-resolved absorption. Proceedings of SPIE - The International Society for Optical Engineering; 2004; 5270, pp. 31–37.
271. Vedrine, C.; Leclerc, J. C.; Durrieu, C.; Tran-Minh, C. Optical whole-cell biosensor using Chlorella vulgaris designed for monitoring herbicides. Biosensors and Bioelectronics 2003, 18, 457–463.
272. Xing, W. L.; Ma, L. R.; Jiang, Z. H.; Cao, F. H.; Jia, M. H. Portable fiber-optic immunosensor for detection of methsulfuron methyl. Talanta 2000, 52, 879–883.
273. Xavier, M. P.; Vallejo, B.; Marazuela, M. D.; Moreno-Bondi, M. C.; Baldini, F.; Falai, A. fiber optic monitoring of carbamate pesticides using porous glass with covalently bound chlorophenol red. Biosensors and Bioelectronics 2000, 14, 895–905.
274. Gao, H. H.; Ayyagari, M. S. R.; Kamtekar, S.; Kumar, J.; Marx, K. A.; Tripathy, S. K.; Kaplan, D. L. Novel fiber-optic biosensor for pesticide detection. Conference Proceedings - Lasers and Electro-Optics Society Annual Meeting; 1994; 8, p. 211.
275. Choi, J. W.; Kim, Y. K.; Lee, I. H.; Min, J.; Lee, W. H. Optical Organophosphorus Biosensor Consisting of acetylcholinesterase/viologen Hetero Langmuir-Blodgett Film. Biosensors and Bioelectronics 2001, 16, 937–943.
276. Choi, J. W.; Kim, Y. K.; Song, S. Y.; Lee, I. H.; Lee, W. H. Optical biosensor consisting of glutathione-S-transferase for detection of captan. Biosensors and Bioelectronics 2003, 18, 1461–1466.
277. Kumar, J.; Jha, S. K.; D'Souza, S. F. Optical microbial biosensor for detection of methyl parathion pesticide using Flavobacterium sp. whole cells adsorbed on glass fiber filters as disposable biocomponent. Biosensors and Bioelectronics 2006, 21, 2100–2105.
278. Svitel, J.; Surugiu, I.; Dzgoev, A.; Ramanathan, K.; Danielsson, B. Functionalized surfaces for optical biosensors: Applications to in

vitro pesticide residual analysis. Journal of Materials Science: Materials in Medicine 2001, 12, 1075–1078.
279. Boiarski, A. A.; Ridgway, R. W.; Busch, J. R.; Turhan-Sayan, G.; Miller, L. S. Integrated optic biosensor for environmental monitoring. Proceedings of SPIE - The International Society for Optical Engineering; 1992; 1587, pp. 114–128.
280. Frense, D.; Muller, A.; Beckmann, D. Detection of environmental pollutants using optical biosensor with immobilized algae cells. Sensors and Actuators, B: Chemical 1998, 51, 256–260.
281. Durrieu, C.; Chouteau, C.; Barthet, L.; Chovelon, J. M.; Tran-Minh, C. A bi-enzymatic whole-cell algal biosensor for monitoring waste water pollutants. Analytical Letters 2004, 37, 1589–1599.
282. Shriver-Lake, L. C.; Kusterbeck, A. W.; Ligler, F. S. Environmental immunosensing at the naval research laboratory. ACS Symposium Series 1996, 646, 46–55.
283. Donner, B. L.; Shriver-Lake, L. C.; McCollum, A., Jr.; Ligler, F. S. Transition from laboratory to on-site environmental monitoring of 2,4,6-trinitrotoluene using a portable fiber optic biosensor. ACS Symposium Series 1996, 657, 197–209.
284. Ives, J. T.; Doss, H. M.; Sullivan, B. J.; Stires, J. C.; Bechtel, J. H. Fiber optic immunosensors to monitor small molecule analytes in groundwater. Proceedings of SPIE - The International Society for Optical Engineering; 1999; 3540, pp. 36–44.
285. Chee, G. J.; Nomura, Y.; Ikebukuro, K.; Karube, I. Optical fiber biosensor for the determination of low biochemical oxygen demand. Biosensors and Bioelectronics 2000, 15, 371–376.
286. Preininger, C.; Kilmant, I.; Wolfbels, O. S. Optical fiber sensor for biological oxygen demand. Analytical Chemistry 1994, 66, 1841–1846.
287. Dai, Y. J.; Lin, L.; Li, P. W.; Chen, X.; Wang, X. R.; Wong, K. Y. Comparison of BOD optical fiber biosensors based on different microorganisms immobilized in ormosil matrixes. International Journal of Environmental Analytical Chemistry 2004, 84, 607–617.
288. Lin, L.; Xiao, L. L.; Huang, S.; Zhao, L.; Cui, J. S.; Wang, X. H.; Chen, X. Novel BOD optical fiber biosensor based on co-immobilized

microorganisms in ormosils matrix. Biosensors and Bioelectronics 2006, 21, 1703–1709.
289. Durrieu, C.; Tran-Minh, C. Optical algal biosensor using alkaline phosphatase for determination of heavy metals. Ecotoxicology and Environmental Safety 2002, 51, 206–209.
290. Hakkila, K.; Green, T.; Leskinen, P.; Ivask, A.; Marks, R.; Virta, M. Detection of bioavailable heavy metals in EILATox-Oregon samples using whole-cell luminescent bacterial sensors in suspension or immobilized onto fibre-optic tips. Journal of Applied Toxicology 2004, 24, 333–342.
291. Kuswandi, B. Simple optical fibre biosensor based on immobilised enzyme for monitoring of trace heavy metal ions. Analytical and Bioanalytical Chemistry 2003, 376, 1104–1110.
292. Leth, S.; Maltoni, S.; Simkus, R.; Mattiasson, B.; Corbisier, P.; Klimant, I.; Wolfbeis, O. S.; Csoregi, E. Engineered bacteria based biosensors for monitoring bioavailable heavy metals. Electroanalysis 2002, 14, 35–42.
293. Lu, Y.; Liu, J.; Li, J.; Bruesehoff, P. J.; Pavot, C. M. B.; Brown, A. K. New highly sensitive and selective catalytic DNA biosensors for metal ions. Biosensors and Bioelectronics 2003, 18, 529–540.
294. Zeng, H. H.; Thompson, R. B.; Maliwal, B. P.; Fones, G. R.; Moffett, J. W.; Fierkel, C. A. Realtime determination of picomolar free Cu(II) in seawater using a fluorescence-based fiber optic biosensor. Analytical Chemistry 2003, 75, 6807–6812.
295. Barker, S.L.R.; Zhao, Y.; Marletta, M.A.; Kopelman, R. Cellular applications of a sensitive and selective fiber-optic nitric oxide biosensor based on a dye-labeled heme domain of soluble guanylate cyclase. Analytical Chemistry 1999, 71, 2071–2075.
296. Mechery, S. J.; Singh, J. P. Fiber optic based gas sensor with nanoporous structure for the selective detection of NO2in air samples. Analytica Chimica Acta 2006, 557, 123–129.
297. Ryoo, D.; Kim, Y.; Jeun, K.; Wood, T.; Choi, Y. A fiber-optic biosensor based on monooxygenases and sol-gel entrapped fluoresceinamine for trichloroethene and tetrachloroethene. 2005 NSTI Nanotechnology Conference and Trade Show - NSTI Nanotech 2005 Technical Proceedings; 2005; pp. 505–507.

298. Shriver-Lake, L. C.; Breslin, K. A.; Golden, J. P.; Judd, L. L.; Choi, J.; Ligler, F. S. Fiber optic biosensor for the detection of TNT. Proceedings of SPIE - The International Society for Optical Engineering; 1995; 2367, pp. 52–58.
299. Shriver-Lake, L. C.; Breslin, K. A.; Charles, P. T.; Conrad, D. W.; Golden, J. P.; Ligler, F. S. Detection of TNT in water using an evanescent wave fiber-optic biosensor. Analytical Chemistry 1995, 67, 2431–2435.
300. Shriver-Lake, L. C.; Donner, B. L.; Ligler, F. S. On-site detection of TNT with a portable fiber optic biosensor. Environmental Science and Technology 1997, 31, 837–841.
301. Shriver-Lake, L. C.; Patterson, C. H.; Van Bergen, S. K. New horizons: Explosive detection in soil extracts with a fiber-optic biosensor. Field Analytical Chemistry and Technology 2000, 4, 239–245.
302. Zhang, S. F.; Wickramasinghe, Y. A. B. D.; Rolfe, P. Investigation of an optical fibre pH sensor with the membrane based on phospholipid copolymer. Biosensors and Bioelectronics 1996, 11, 11–16.

CHAPTER 6

Single- and Two-Phase Flow Characterization Using Optical Fiber Bragg Gratings

Virgínia H.V. Baroncini [1], Cicero Martelli [1,*], Marco José da Silva [1] and Rigoberto E.M. Morales [2]

[1] Graduate School of Electrical Engineering and Computer Science, Federal University of Technology-Paraná (UTFPR), Curitiba 80230-901, Brazil

[2] Graduate School of Mechanical and Materials Engineering, Federal University of Technology-Paraná (UTFPR), Curitiba 80230-901, Brazil

ABSTRACT

Single- and two-phase flow characterization using optical fiber Bragg gratings (FBGs) is presented. The sensor unit consists of the optical fiber Bragg grating positioned transversely to the flow and fixed in the pipe walls. The hydrodynamic pressure applied by the liquid or air/liquid flow to the optical fiber induces deformation that can be detected by the FBG. Given that the applied pressure is directly related to the mass flow, it is possible to establish a relationship using the grating resonance wavelength shift to determine the mass flow when the flow velocity is well known. For two phase flows of air and liquid, there is a significant change in the force applied to the fiber that accounts for the very distinct densities of these substances. As a consequence, the optical fiber deformation and the correspondent grating wavelength shift as a function of the flow will be very different for an air bubble or a liquid slug, allowing their detection as they flow through the pipe. A quasi-distributed sensing tool with 18 sensors evenly spread along the pipe is developed and characterized, making possible the characterization of the flow, as well as the tracking of the bubbles over a large section of the test bed. Results show good agreement with standard measurement methods and open up plenty of opportunities to both laboratory measurement tools and field applications.

KEYWORDS

two-phase flow; single-phase flow; strain sensor; optical fiber Bragg grating

1. INTRODUCTION

The simultaneous stream of liquid and gas in a pipe, known as gas-liquid two-phase flows, occurs in many industrial applications, such as chemical, oil and nuclear industries. These flows have been an important field of study for physics and mechanical engineering. Therefore, there has been much effort given by the scientific community and industry for the development of monitoring solutions for such conditions, since they are very common and have a large economic impact regarding equipment usage, as well as production efficiency [1]. Two main issues limit the understanding of real-world large-scale problems. First is the difficulty of scaling down flow problems into small laboratory models. For instance, if a several kilometers-long pipeline that is 17" in diameter is transformed into a laboratory-scale pipeline in terms of length, the diameter of the tubes would be on the order of microns or less, unless the density of the fluid is changed, which is also extremely complicated when one is considering the scale factor of kilometers down to meters. Secondly, the computational demand for solving fundamental equations of real-world-sized systems is extremely large, and most of the time, it just cannot be done. Although some researchers have produced successful solutions to engineering two-phase flow problems, the accuracy limits and the general applicability of simulations depend intrinsically on empirical relationships obtained in experiments. Several solutions based on capacitive/resistive sensors [2], gamma-ray densitometry [3], particle imaging velocimetry [4] and wire-mesh sensor technology [5] have been presented, and some are commercially available.

Optical fiber sensors make use of many of the advantages that come from silica optical fibers, which have made this technology suitable for a wide range of environment applications. They are typically small in size, electrically and chemically passive, immune to electromagnetic interference, robust to harsh environments and have a capability to perform high resolution distributed sensing. They have the down side, however, of the need to be in direct contact with the physical quantity been measured, which sometimes turns them into an intrusive sensor. For this specific application, the sensor is intrusive, but it impacts the measurement or the monitored phenomena is negligible, provided its quite small diameter. Given that optical fibers were first developed for communication systems, optical fiber sensors benefit from the important possibility of having links in communications, such as high, dense multiplexing, and so on. Optical fiber Bragg grating sensors are increasingly being used in specific applications in aeronautics, the automotive industry, structure monitoring in civil engineering and undersea oil exploration [6,7,8]. A range of parameters are being measured, such as strain [9], force [10], pressure [11], displacement [12], temperature [13], humidity [14] and radiation dose [15]. For flow measurement, some researchers have applied optical fiber sensors for flow rate measurement in the past [16,17,18], which were basically limited to single-phase flows. More recently, researchers have presented studies employing FBGs for two-phase flow detection in microchannels [19], as an anemometer with a silver-coated FBG [20] and for the measurement of flow the rate and direction using a cantilever [21].

In this paper, an alternative technique is proposed using optical fiber Bragg grating strain sensors to detect single- and two-phase flow problems. The optical fiber gratings are placed transversally to the flow and are subjected to the forces originated by the flowing fluid. As the liquid flows through the optical fiber, it induces a mechanical force along the fiber, which is proportional to the hydrodynamic pressure and, consequently, to liquid flowing mass. As the fiber is subjected to the flow-induced force, the optical fiber Bragg grating wavelength shifts proportionally. For the single-phase flows, the flow velocity is determined basically by the strain measured by the gratings, since the fluid density is constant and well known. In the case of a two phase flow, the speed of the flow is determined by using two gratings positioned at fixed distances between each other, as demonstrated below in the paper. As this technique is based on the property of the flowing fluid, it is possible that the two-phase flow might be characterized by a smaller number of sensors than is necessary with conventional techniques. The capacity of multiplexing tens to hundreds of optical fiber sensors along the pipe is also another advantage of the proposed technique. This allows the characterization of the flow properties, including air bubble and liquid slug velocity, size and shape and their changes as they flow over the pipe length.

2. SENSOR UNIT AND EXPERIMENTAL SETUP

The concept of the sensing element consists of an optical fiber Bragg grating positioned transversely to the flow and fixed in the pipe walls. After several experimental tests, it was concluded that the best assembly condition, as well as the best sensor response were achieved when the grating was glued to the pipe walls without any tension. As the fluid flows around the grating, it shifts its wavelength, and the dynamic pressure of the flow is detected. The strain induced in the optical fiber is also a resultant effect of bending displacement and vibration, as shown in Figure 1a,d.

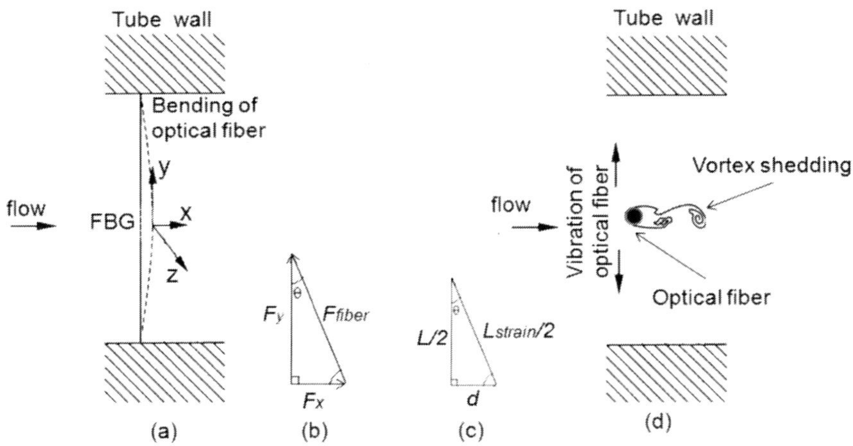

Figure 1. Schematic drawing of the forces that appear in the optical fiber inserted across a flow: (**a**) side view; (**b**) force diagram; (**c**) diagram showing the changes in dimension of the strained optical fiber; and (**d**) top view of the acting forces. FBG, fiber Bragg grating.

The frequency of such vibration depends on the force exerted by the flow onto the fiber optics and also the density of the fluid, i.e., it depends on the local Reynold's number. The force field acting on the optical fiber is transverse at the fiber, resulting in changes in the Bragg wavelength. These changes will allow the determination of the force of the flow acting on the fiber. The force field in the horizontal direction (x-axis) is obtained by the shift of fiber Bragg gratings in the optical fiber. As a matter of simplification, one can consider that the entire force field generated by the flow over the optical fiber transverse to the flow can be represented by a single force component represented by *Fx*.

When an optical fiber is stretched, it suffers a strain, which is proportional to the amplitude of the longitudinal force Fy = εEA along the fiber length, where E is the Young's modulus of silica fiber and *A* is the cross-sectional area of the fiber. For an optical fiber subjected to a force *Fx* (see Figure 1b) in the x-direction, a force component on the y-direction also appears, which leads to strain along the fiber. This strain can be measured by the wavelength shift of an FBG recorded in the optical fiber. Therefore, one can use such a wavelength shift to determine the force *Fx* acting on the fiber. Decomposing the force diagram (Figure 1b):

$$F_y = F_{fiber} cos\theta \quad \text{and} \qquad F_x = F_y tg\theta \tag{1}$$

where *FFiber* is the force on the fiber. Considering that θ is very small, it is possible to conclude that *Fy=Ffiber*. In Figure 1c, an approximation of the geometric changes in the fiber as the flow pushes the fiber is presented. L/2 and $L_{strain}/2$ correspond to half of the strained fiber and half of the strained fiber, respectively, and d is the displacement of the fiber from its unstrained state to the strained position. Considering that:

$$tg\theta = \frac{d}{L/2} \tag{2}$$

$$L_{strain} = L(\varepsilon + 1) \tag{3}$$

$$\varepsilon = \Delta L/L \tag{4}$$

It is possible to determine that:

$$tg\theta = \sqrt{(\varepsilon + 1)^2 - 1} \tag{5}$$

Additionally, the force acting on the fiber is:

$$F_x = \varepsilon . E . A\sqrt{(\varepsilon + 1)^2 - 1} \tag{6}$$

Given that the Bragg wavelength shift as a result of the applied strain, this can be expressed by [20]:

$$\Delta\lambda_B = \lambda_B\left(1 - \rho_{eff}\right)\varepsilon \tag{7}$$

where λB is the FBG Bragg wavelength, $\Delta\lambda B$ is the wavelength shift and ρeff is an effective strain-optic coefficient [22]. The amplitude of the force field in the horizontal direction (x-axis) obtained by the shift of fiber Bragg gratings in the optical fiber is determined by:

$$F_x = \left(\frac{\Delta\lambda_B}{\lambda_B\left(1 - \rho_{eff}\right)}\sqrt{\left(\frac{\Delta\lambda_B}{\lambda_B\left(1 - \rho_{eff}\right)} + 1\right)^2 - 1}\right).E.A \tag{8}$$

The mass flow (m) can also be derived from the equations above, since it is proportional to the force (Fx), as shown below:

$$m \approx \frac{F_x}{K.v^2} \tag{9}$$

where v is the flow velocity and K is the constant of proportionality, which depends on the optical fiber sensor assembly properties.

The sensor setup using optical fiber Bragg gratings is shown in Figure 2b. The fiber Bragg gratings were fabricated using standard single-mode optical fiber (SMF), which was subjected to the 193-nm radiation of an Excimer laser through a phase mask. The spectral response of the FBGs was monitored during the fabrication in order to assure that similar reflectivities for all gratings were achieved. In order to position the FBGs transversally to the flow, metallic plates with V-groove-like channels were fabricated and used to support the gratings, as is shown in Figure 2a. Eighteen gratings multiplexed in wavelength are used in order to allow the monitoring of the flow properties along the length of the test bed.

The sensor unit is connected to an 8 m-long arm of a fluid flow simulation setup (see Figure 2c). A rotating pump is used to generate water flow with rates ranging from 0 to 2.5 m/s, while air from a compressor is mixed with the flowing water, generating the two-phase flow. Water and air flows are both independently computer controlled in order to obtain different flow regimes. The FBG wavelengths were monitored using a commercially-available dynamic interrogator (I-MON 512 USB) and the broadband emission of a super luminescent LED source (DL-BP1-1501A SLED) from Ibsen Photonics Inc., as is schematically represented in Figure 2b. The interrogator was operated with a repetition rate of 800 Hz.

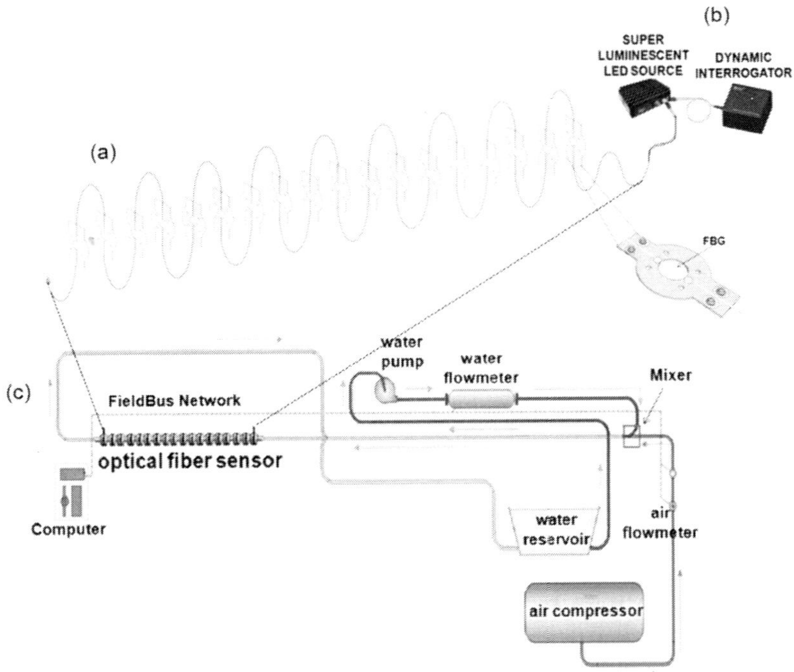

Figure 2. (**a**) Schematic drawing of the optical fiber strain sensors installed in a pipe; (**b**) Measuring setup; (**c**) Experimental rig for simulating different flow regimes.

3. RESULTS AND DISCUSSION

As introduced above, the monitoring of single- and two-phase flows is of great importance to both academic and industrial applications. The latter is because of the ever growing need for controlling and supervising production plants. The first is related to the lack of understanding of many distinct flow conditions, as well as the limitations of the mathematical simulation tools. Experimental measurements can be used to produce design relations, through simple mathematical correlations and establish relationships between the measured variables. In the following, the sessions results of the proposed sensing technique will be presented for single-phase flow of water and a two-phase flow of water and air.

3.1. Single-Phase Flow Measurements

In Figure 3, the measurement of single-phase flows for different velocities is presented. Figure 3a shows the effect of the flow velocity on the optical fiber sensor. The time series corresponds to a measurement where the velocity for every segment of the plot is kept constant for approximately 5 s, and it is increased in steps of 0.2 m/s, in order to observe the increase in the signal oscillation as a function of the local Reynold's number. Two phenomena are observed: (1) the quadratic dependence of the strain against the velocity; (2) an increase of the signal noise, which results from the rising of the local Reynold's number. Figure 3b shows the effect of the force created by the fluid at distinct speeds resulting in the Bragg wavelength shift on the optical fiber.

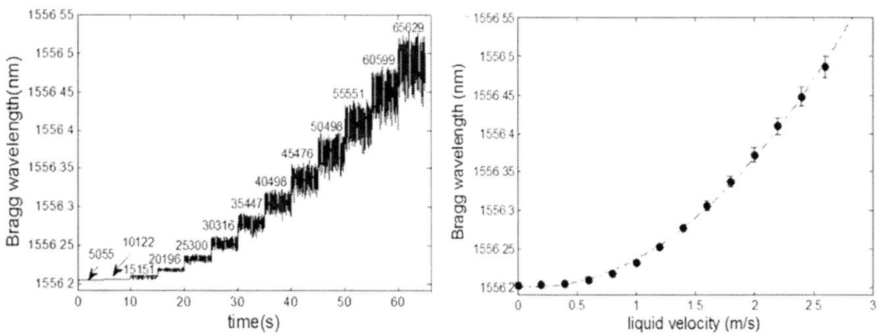

(a) (b)

Figure 3. (**a**) Behavior of an FBG strain for single-flow and increasing velocities; the Reynold's number increases, leading to increasing high frequency oscillations induced on the optical fiber sensor; (**b**)Typical optical fiber Bragg grating wavelength shift and standard deviation against single-phase flow liquid velocity.

This means that it is possible to measure both the speed of the flow, as well as the mass that is flowing through the pipe once the density of the fluid is known. To find the velocity of the liquid measured by the FBG, one can determine the inverse function of the quadratic relationship between the Bragg wavelength and the fluid velocity by the flowmeter.

The comparison between the FBG measured mass flow against the value determined by the standard sensor presented in the test bed is shown in Figure 4. The reference sensor consists of a commercially-available Coriolis sensor, model TCM-230K, manufactured by Tricor, which determines the water flow that can further be transformed into mass flow by using the water density.

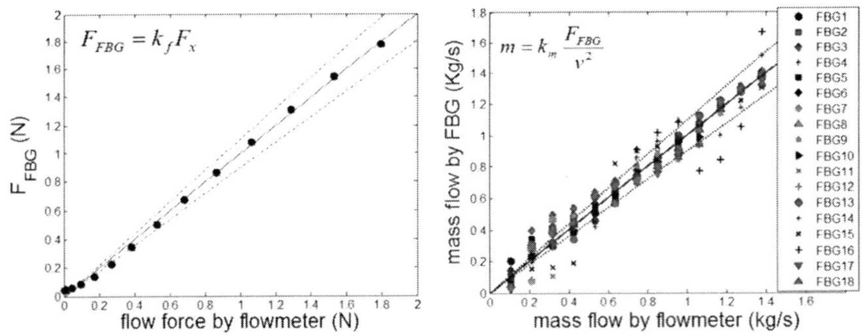

(a) (b)

Figure 4. (**a**) Measurement of the flow-induced force to the optical fiber Bragg grating against the flow force determined by the flow properties; (**b**) Measurement of the mass flow for all 18 sensors, indicating the good agreement between the FBG and standard mass flow meter. In both plots, the solid line is at 45°, indicating the position where all values should converge, and the dashed lines indicate a ±10% error.

The performance of the 18 sensors distributed along the pipe against a reference sensor is shown in Figure 4, and overall, all of the sensors operate with the minimum deviation. Small differences are noted in

Sensors 15 and 16, for instance, which are due to problems in the manufacturing, including gluing and alignment.

3.2. Two-Phase Flow Measurements

The most common flow patterns in horizontal flows are known as dispersed bubble, slug flow, stratified flow and annular flow [1]. Among these, the slug flow occurs in a wide range of flow rates of liquid and gas. The slug flow is characterized by the succession of two distinct regions: the slug, which consists of a liquid column, and an elongated bubble, as shown in Figure 5a.

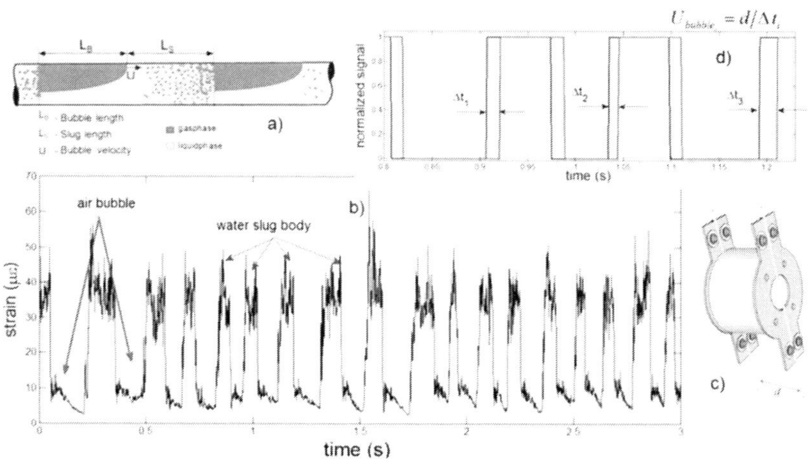

Figure 5. (**a**) Schematic representation of the slug flow with detailed information about the air bubble and liquid slug length and shape; (**b**) Example of a time series detected by the FBG for a typical slug flow for water and air velocities of 2 m/s; (**c**) Detail of the sensor assembly showing a pair of sensors and the fixed distance d = 5 cm between them; (**d**) Treated signal for two FBGs as in (c), indicating the time lag Δt between the measured signals that is used to measure the bubble velocity.

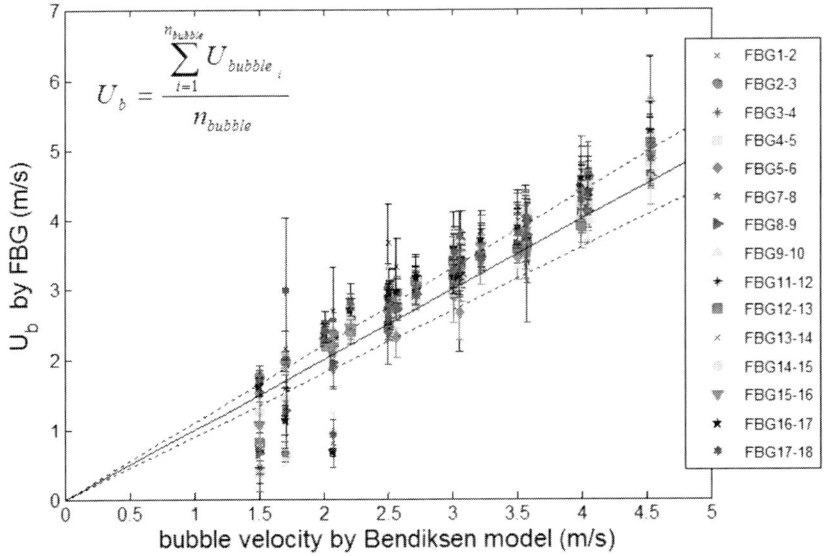

Figure 6. Bubble velocity measurement, using 18 fiber Bragg gratings sensors.

The raw data measured by the optical fiber Bragg grating sensors is shown in Figure 5b. When there is only water passing through the optical fiber sensor, the signal amplitude indicates a maximum load being applied to the fiber. The presence of a bubble decreases the volume of water in the tube cross-section, subsequently decreasing the amplitude of the forces on the fiber sensor, which results in the relaxation of the strain in the fiber, reaching a minimum value. The noisy-like oscillation observed at the higher strain values is possibly originated from the turbulence in the flow. This oscillation is the same as observed in Figure 3a, where the flow velocity increases, as well as the Reynold's number and, thus, the oscillation induced on the optical fiber sensor. The comparison for all distinct liquid and gas speeds is presented in Figure 6. The x-axis indicates the theoretical expected values for the bubble speed considering Bendiksen's model, which takes into account the liquid and the gas flow injected into the main pipe of the test bed [23].

4. CONCLUSIONS

This paper has covered a novel sensing technique to characterize single and two-phase flows. The use of FBG strain sensors as an alternative solution to directly measure the mass flow of the single-phase flow of liquids in pipes is demonstrated. The technique proposed here can be used with virtually any liquid. The operating mechanism of the sensors is based on the measurement of the strain induced by the flowing fluid over the optical fiber. It was also shown that optical fiber sensors based on optical fiber Bragg gratings are particularly suitable for applications where the characterization of the bubble or slug profiles over the pipe length is needed. The sensing technique demonstrated here brings new opportunities for the detection and characterization of single- and dual-phase flows, and it is envisaged that with the correct packaging, this sensor can readily be implemented in real-world problems, where small, intrusive sensors can be used. This includes all sorts of small diameter pipelines present in the chemical sector, for instance. The optical fiber Bragg grating sensors consist, therefore, of an interesting technology, which is minimally intrusive and can be used to generate various information about the sensor with a minimum number of sensors.

ACKNOWLEDGMENTS

The authors acknowledge the financial support of CNPq (Brazilian National Research Council), Petrobras, Fundação Araucaria and FINEP (Brazilian government company devoted to finance and fund innovation and research).

AUTHOR CONTRIBUTIONS

Virgínia H.V. Baroncini, Cicero Martelli, Marco José da Silva and Rigoberto E.M. Morales conceived and designed the experiments; Virgínia H.V. Baroncini performed the experiments; Virgínia H.V. Baroncini, Cicero Martelli, Marco José da Silva and Rigoberto E.M. Morales analyzed the data; Virgínia H.V. Baroncini wrote the paper.

CONFLICTS OF INTEREST

The authors declare no conflict of interest.

REFERENCES

1. Azzopardi, B. Gas-Liquid Flows; Begell House: New York, NY, USA, 2006.

2. Da Silva, M.J.; Schleicher, E.; Hampel, U. Capacitance wire-mesh sensor for fast measurement of phase fraction distributions. Meas. Sci. Technol. 2007, 18, 2245–2251.

3. Johansen, G.A. Nuclear tomography methods in industry. Nuclear Phys. A 2005, 752, 696–705.

4. Hassan, Y.A. Multiphase bubbly flow visualization using particle image velocimetry. Ann. N. Y. Acad. Sci. 2002, 972, 223–228.

5. Da Silva, M.J.; Thiele, S.; Abdulkareem, L.; Azzopardi, B.J.; Hampel, U. High gas-oil two flow visualization with a capacitance wire-mesh sensor. Flow Meas. Instrum. 2010, 21, 191–197.

6. Kersey, A.; Davis, M.; Patrick, H.; LeBlanc, M.; Koo, K.P.; Putnam, M.; Firebele, E. Fiber grating sensors. J. Lightwave Technol. 1997, 15, 1442–1463.

7. Cheng, S.; Zhang, Y. Application of Fiber grating sensor for oil leak detection in oil industry. Appl. Opt. 2008, 29, 441–443.

8. Hassan, M.; Tamchek, N.; Abas, A.; Johar, R.; Adikan, F. Dual-phase sensing for early detection of prepreg structural failures via etched cladding Bragg grating. Sens. Actuator A Phys. 2011, 171, 126–130.

9. Schizas, C.; Stutz, S.; Botsis, J.; Coric, D. Monitoring of non-homogeneous strains in composites with embedded wavelenght multiplexed fiber Bragg gratings: A methodological study. Compos. Struct. 2012, 94, 987–994.

10. Zhao, Yo.; Zhao, Ya.; Zhao, M. Novel force sensor based on a couple of fiber Bragg gratings. Measurement 2005, 38, 30–33.

11. Urban, F.K.J.; Vlach, R.; Kuchta, R. Design of a pressure sensor based on optical fiber Bragg grating lateral deformation. Sensors 2010, 10, 11212–11225.

12. Ng, J.; Zhou, X.; Ang, X.Y.; Hao, J. A simple temperature-insensitive fiber Bragg grating displacement sensor. Opt. Commun. 2007, 273, 398–401.

13. Zhan, Y.; Xue, S.; Yang, Q.; Xiang, S.; He, H.; Zhu, R. A novel fiber Bragg grating high-temperature sensor. Optik 2008, 119, 535–539.

14. Yeo, T.; Sun, T.; Grattan, K.; Parry, D.; Lade, R.; Rowell, B.D. Characterisation of polymer-coated fibre Bragg grating sensor for relative humidity sensing. Sens. Actuator B Chem. 2005, 110, 148–156.

15. Fernandez, A.; Brichard, B.; Berghmans, F.; Decreton, M. Dose-rate dependencies in gamma-irradiated fiber Bragg grating filters. IEEE Trans. Nucl. Sci. 2002, 49, 2874–2878.

16. Barton, J.S.; Saoudi, M. A fiber optic vortex flowmeter. J. Phys. E Sci. Instrum. 1986, 19, 64–66.

17. Takashima, S.; Asanuma, H.; Niitsuma, H. A water flowmeter using dual fiber Bragg grating sensors and cross-correlation technique. Sens. Actuators A Phys. 2004, 116, 66–74.

18. Zhao, Y.; Chen, K.; Yang, J. Novel target type flowmeter based on a differential fiber Bragg grating sensor. Measurement 2005, 38, 230–235.

19. Nigel, D.; Djilali, N.; Wild, P. Fiber Bragg grating sensor for two-phase flow in microchannels. Microfluid. Nanofluid. 2012, 13, 99–106.

20. Lu, P.; Chen, Q. Fiber Bragg grating sensor for simultaneous measurement of flow rate and direction. Meas. Sci. Technol. 2008, 19.

21. Wang, X.; Dong, X.; Zhou, Y.; Li, Y.; Cheng, J.; Chen, Z. Optical fiber anemometer using silver-coated fiber Bragg grating and bitaper. Sens. Actuators A Phys. 2014, 214, 230–233.

22. Othonos, A. Fiber Bragg gratings Fundamentals and Applications in telecommunications and Sensing; Artech House, Inc.: Norwood, MA, USA, 1999.

23. Bendiksen, K.H. An experimental investigation of the motion of long bubbles in inclined tubes. Multiph. Flow 1984, 10, 467–483.

CHAPTER 7

Design and Deployment of Low-Cost Plastic Optical Fiber Sensors for Gas Monitoring

Sabrina Grassini [1,*], Maen Ishtaiwi [2], Marco Parvis [2] and Alberto Vallan [2]

[1] Department of Applied Science and Technology, Politecnico di Torino, Corso Duca degli Abruzzi, 24 Torino, Italy

[2] Department of Electronic and Telecommunications, Politecnico di Torino, Corso Duca degli Abruzzi, 24 Torino, Italy

ABSTRACT

This paper describes an approach to develop and deploy low-cost plastic optical fiber sensors suitable for measuring low concentrations of pollutants in the atmosphere. The sensors are designed by depositing onto the exposed core of a plastic fiber thin films of sensitive compounds via either plasma sputtering or via plasma-enhanced chemical vapor deposition (PECVD). The interaction between the deposited layer and the gas alters the fiber's capability to transmit the light, so that the sensor can simply be realized with a few centimeters of fiber, an LED and a photodiode. Sensors arranged in this way exhibit several advantages in comparison to electrochemical and optical conventional sensors; in particular, they have an extremely low cost and can be easily designed to have an integral, i.e., cumulative, response. The paper describes the sensor design, the preparation procedure and two examples of sensor prototypes that exploit a cumulative response. One sensor is designed for monitoring indoor atmospheres for cultural heritage applications and the other for detecting the presence of particular gas species inside the RPC (resistive plate chamber) muon detector of the Compact Muon Solenoid (CMS) experiment at CERN in Geneva.

KEYWORDS

gas sensors; plasma deposition; PECDV; plastic optical fibers

1. INTRODUCTION

Fiber optic sensors are being employed for various sensing applications, e.g., optical fiber sensors are used for detecting the presence of air pollutants in the atmosphere [1,2], as well as being employed in the biomedical field [3,4] and in environmental sensing [5,6]. This wide diffusion is mainly due to the relevant advantages of optical fibers, such as the intrinsic fire-safe behavior and the immunity to electromagnetic interference (EMI), which make them suitable for in situ measurements, also in adverse environments.

The most common approach to obtain a fiber optic sensor (FOS) is to use a glass fiber, whose core is modified to have a periodic change of its refractive index, thus creating the so-called fiber Bragg grating (FBG). FBG sensors can be employed to arrange intrinsic strain and temperature sensors and are widely used, in conjunction with assemblies that produce controlled fiber strain, to measure different kinds of quantities. The main problem of this kind of sensor is the cost of the interrogation system, which must be capable of selectively measuring at specific wavelengths. In addition, the small diameter of typical glass fibers requires special attention in their illumination and the use of low-tolerance, high-cost connectors.

Plastic optical fibers (POF), in comparison, have a lower cost per meter and usually a higher diameter that results in relaxed specifications on the connectors and in easy illumination. On the other hand, POF have a higher attenuation with respect to glass fibers; this limits their use in the transmission field, but is usually not a problem for sensor development, so that POF use in this field is continuously increasing [7].

POF can be used to arrange sensors by using most of the approaches already employed in glass fibers. FBG, both normal and long-period, can be inscribed into low-cost plastic optical fiber to obtain different kinds of sensors [8], but the problem of the interrogation system cost still may limit the possible applications. Similarly, other proposed solutions, such as backscattering, optical time-domain reflectometers, fluorescence measurements [9] and selective excitation of surface plasma waves in the so-called surface plasmon resonance [10], even though capable of providing quite interesting results, usually require complex and costly setups.

On the other end, solutions based on the fiber light attenuation measurements could be extremely cheap and simple [11], even though at the expense of reduced stability, as the fiber light transmission capability is affected by a wide number of parameters.

By measuring the fiber light attenuation, it is very easy to directly arrange bending and pressure sensors, but the measurement of chemical quantities requires a chemical modification of the fiber core surface in order to make it reactive with respect to different chemical species with high sensitivity and selectivity [7].

This latter solution has great potentialities, and it can be employed for many different quantities simply by changing the nature of the sensitive coating. Sensors have been developed for measuring the pH of liquid solutions and for detecting the presence of several chemicals, such as methanol, ethanol, toluene and other gases in the air [7]. All of these sensors are designed to detect the amount of chemical present, but the use of specific coatings opens the possibility of designing sensors having a cumulative response, i.e., of devices whose output is connected with the total exposition of the sensor to a specific quantity. Designing sensors with this behavior requires the development of coatings that are affected by non-reversible chemical reactions, but that enable the creation of sensors capable of detecting the presence of extremely low concentrations of specific pollutants

This paper, therefore, describes a procedure that has been successfully employed by the authors to arrange simple and inexpensive plastic optic-based sensors that exploit a cumulative response. These sensors are designed to detect chemicals in the gas phase and to exploit the interaction of the chemicals with the evanescent electromagnetic field at the core surface. The sensor assemblies employ a light amplitude measurement, which requires only an LED and a photo-diode and are arranged by taking advantage of vacuum deposition techniques to make the fiber sensitive to the quantity of interest.

2. PLASTIC OPTICAL FIBER SENSORS

As discussed in the previous section, in order to deploy a fiber optic-based sensor whose transmittance is affected by the materials surrounding the fiber itself, two operations have to be performed.

Firstly, the fiber core must be exposed to place the evanescent field in contact with the surroundings. This is done by removing the fiber jacket and then by removing the fiber cladding, thus exposing the fiber core (Figure 1).

Secondly, a sensitive material, often referred to as modified cladding, has to be chosen and deposed onto the core in such a way that the presence of the substance to be detected is capable of changing the fiber optical properties, thus affecting the fiber transmittance (Figure 2).

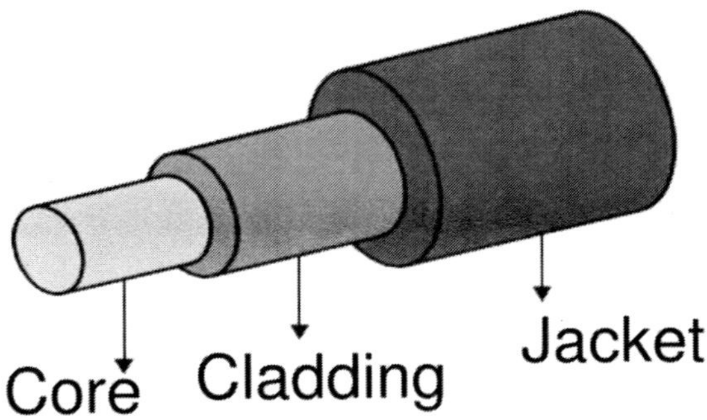

Figure 1. Plastic optical fiber structure (not to scale).

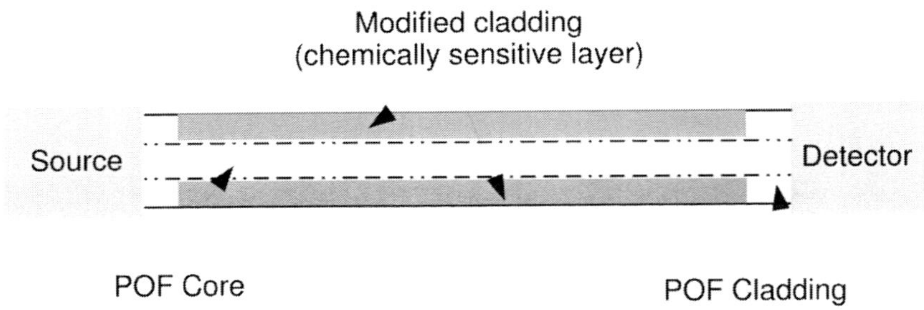

Figure 2. Configuration of the plastic optical fiber sensor.

Several kinds of plastic fibers can be found on the market; however, the most common and inexpensive fibers are the ones designed for short-range telecommunication connections, which have costs of the order of 1€per meter. These fibers are highly multi-mode with a typical diameter of 1 mm, with a poly-methyl-methacrylate (PMMA) core diameter of about 980 μm and a fluoropolymeric cladding 10 μm thick.

The first step is therefore the removal of the fluoropolymeric cladding, which has to be obtained without damaging the core, a condition that could easily lead to unacceptable transmission losses and or a fragile core. Unfortunately, the cladding has mechanical properties quite similar to the core, so a mechanical approach is usually unsuitable, as it is difficult to discriminate when the cladding removal is complete.

A chemical approach, by using a solvent capable of attacking the cladding without damaging the core, is a viable alternative, since the cladding of most POF is composed of a fluoropolymer, while the core is PMMA. As an example, Merchant [12] describes a solution based on the use of acetone, which has been successfully applied to uncover the core. The authors instead decided to use an approach based on ethyl-acetate, which can be easily applied also in the absence of specific tools. The procedure consists of dipping the fiber in ethyl-acetate until the solvent breaks the bonds between the cladding and the PMMA, so that the cladding can be easily rubbed away by using a simple tissue-paper without damaging the core.

The exposure time is critical, since too long of a dipping in the solvent can damage the PMMA core; however, consistent results in the case of typical fibers with a 1 mm diameter can be obtained by using 99.5% ethyl-acetate for 40 s at a temperature of about 25 °C.

The effectiveness of the cladding removal can be assessed by using a field emission scanning electron microscope (FESEM) to observe the core.

Figure 3 shows two images obtained with a Supra 40 Zeiss FESEM equipped with an energy dispersive X-ray spectrometer (EDS) for elemental analysis. The pictures show how the proposed procedure is able to completely remove the cladding. The PMMA core presents only a few small defects with a size of less than 100 nm and a thickness of a few nanometers without any other visible damage.

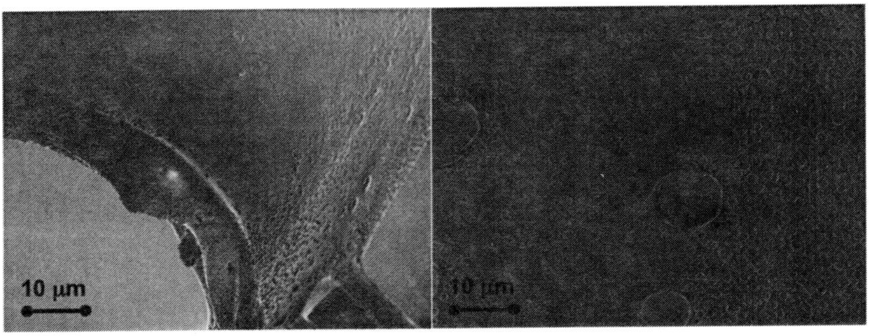

Figure 3. The effect of the ethyl-acetate procedure on the fiber surface. (**Left**) The point where the cladding is removed: the 10 μm edge is clearly visible; (**Right**) The core surface after the removal procedure: the core appears clean and undamaged.

Once the core is exposed, it is possible to coat it by using several different techniques, depending on the composition of the modified cladding, with the only limitation connected with the limited thermal resistance of the plastic fiber. If, as usual, the fiber is made of PMMA, the maximum temperature is typically limited to 70 °C. Among the different techniques, sol gel [13] has been used for the deposition of thick coatings, which are sensitive to pH and other quantities mainly in the liquid phase, while vacuum techniques, such as DC plasma sputtering and plasma-enhanced chemical vapor deposition (PECVD) [14] have been used for making sensors sensitive to gas compounds.

Sections 3 and 4 describe two examples of procedures for the development of gas sensors, which are based on DC plasma sputtering and PECVD, but, as described above, other procedures can be used, as well.

Basically, two approaches can be followed to select the coating material. One possibility is to use a material that initially reflects the light, i.e., which does not adsorb the evanescent field and which becomes less reflective in the presence of the substance to be detected. This approach results in a sensor that has a decreasing transmittance when the substance to be detected is present. On the other hand, if the modified cladding makes the fiber lossy, usually as a consequence of a high refractive index that decreases when the substance to be detected is present, the sensor has an increasing transmittance when the substance is present. In the following sections, two examples referring to the two types of sensors will be described along with their deployment and characterization.

Eventually, the sensitized fiber is interrogated by using a light source and a light-sensing device and has to be mounted, so that its mechanical configuration remains stable. This last point is extremely important, since the transmittance of highly multi-mode fibers strongly depends on fiber bending [15,16].

To this aim, an arrangement, like the one shown in Figure 4, can be used. The assembly is composed of a piece of fiber with a length of the order of 10 cm, an LED, a photodiode and a PMMA support, which holds the fiber and diodes. Both the LED and photodiode have the conventional 5 mm diameter case. These semiconductors are prepared by drilling a 1 mm hole through their top: a PMMA glue is used to bond the fiber, whose ends are inserted into the holes, to the diodes. The assembly is then cured by putting it into an oven at 60 °C for four hours to polymerize the glue. Eventually, the assembly is fixed onto the support.

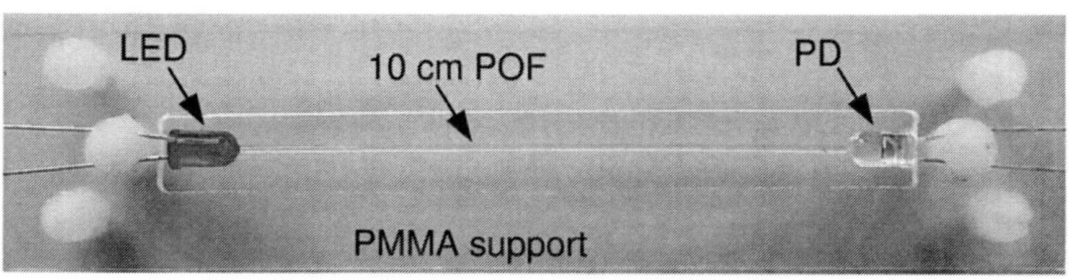

Figure 4. Sensor assembly composed of fiber, LED and photodiode.

By using this assembly, most of the problems connected with either mechanical bending or thermal dilatation are made negligible, since the thermal expansion coefficients of both the fiber and support are quite similar. In addition, the mounting procedure with plastic screws allows one to easily mount and unmount the fiber and, therefore, to coat the core, either before or after having it mounted on the holder. After the sensor is arranged, i.e., after the fiber core is coated with the sensitive material, the photodiode has to be painted, to avoid capturing external light.

Figure 4 shows the use of a red LED, which has been selected in this prototype according to the considerations expressed in Section 4, but other LED colors can be used, as well, depending on the specific behavior of the used coating.

3. H_2S SENSOR OBTAINED VIA DC PLASMA SPUTTERING EXAMPLE

This first example describes a solution designed for the cultural heritage field [17], where old wood showcases can produce small amounts of hydrogen sulfide (H_2S) and of formic acid (HCOOH). Both are usually generated in less than a part-per-million concentration, thus being hardly detectable by conventional sensors, but the exposition of metallic artifacts, e.g., silver jewels and coins, to them for a long time leads to tarnishing and impairs their conservation and aspect, so that it should be avoided. A fiber-based sensor with a cumulative response in this case can act as an extremely sensitive device capable of detecting the overall exposure to the gas compounds instead of the gas concentration, which has a very low value. This behavior can be easily obtained by employing silver as the sensitive layer. The gas reaction of silver with hydrogen sulfide is:

$$2Ag^{+} + S^{--} \Rightarrow Ag_2S + 2e^{-} \tag{1}$$

Such a reaction is not reversible and has a first-order law kinetic with respect to the sulfide concentration S_C:

$$-\frac{dS_{Ag}}{dt} = k_S S_C \tag{2}$$

where k_S is the kinetic constant that mainly depends on the temperature and S_C is the sulfide concentration.

The effect of this tarnishing reaction is that the silver surface loses its brilliance with time, becoming darker. The idea is therefore to coat the fiber core with a very thin layer of silver, so that any tarnishing reaction occurs very quickly and to observe the transmittance reduction due to the silver reflectivity loss as the metallic silver turns into silver sulfide.

Since silver is metallic, a simple DC sputtering, like the one shown in Figure 5, can be used to deposit it onto the fiber.

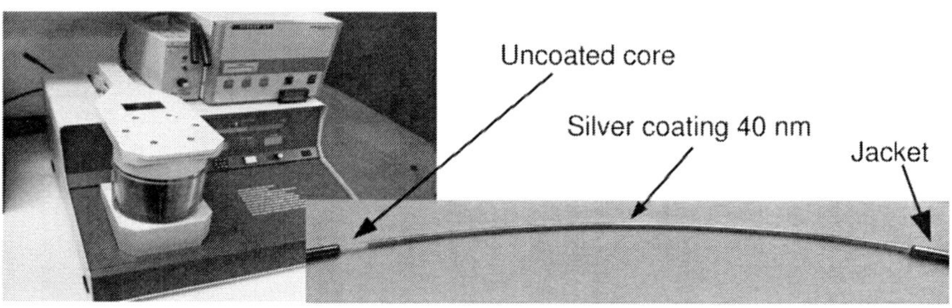

Figure 5. DC sputter used for the deposition (**left**) and an example of a fiber coated with 40 nm of silver (**right**).

The sensing film must have a thickness that is enough to interact with the evanescent field, but small enough to be quickly affected by the tarnishing reaction.

This way, the tarnishing reaction, which is non-reversible, is able to produce measurable propagation losses when the sensor exposed is to the sulfide vapors. After some preliminary tests, a value of the order of 40 nm was found to provide the best values. The DC plasma reactor employed an Ag target (purity 99.99%) and was run by using Ar (purity 99.99%) as the discharge gas; the pressure was taken to a few Pascals by using a turbo-molecular pump in order to obtain a nanostructured silver layer.

Figure 5 also shows a picture of a fiber coated with a 40 nm-thick silver layer, which is used to arrange the sensor assembly as described in previous section.

Since the sensor is designed to detect very low concentrations of H_2S, a direct characterization of it by using a standard sensor is quite difficult to obtain, and in addition, it is also difficult to provide stable and very low concentrations of H_2S vapor. The solution adopted by the authors therefore is a characterization carried out by putting the sensor inside a reaction chamber, where, from time to time, a small amount of gas containing H_2S is injected. The chamber is not sealed on purpose, so that, after a certain time, the air circulation removes all of the H_2S, back to clean air. Figure 6 shows the proposed system, which employs a standard electrochemical H_2S sensor designed for a range of 50 ppm as the reference measuring system.

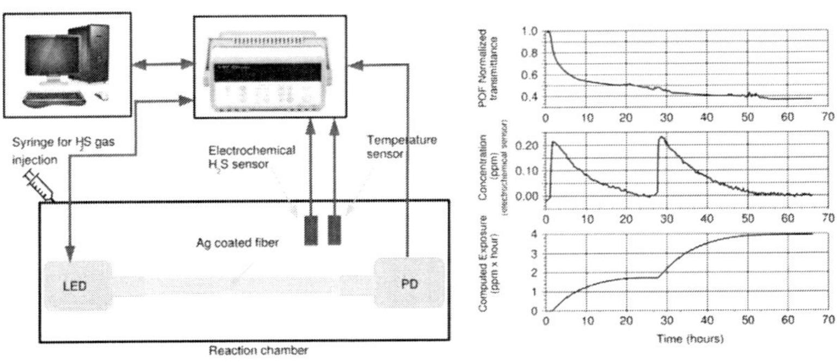

Figure 6. Setup used for characterizing the H_2S sensor (**left**) and an example of traces obtained during the test (**right**).

In order to obtain meaningful measurements in the presence of very small H_2S concentrations, the zero level of the reference sensor is compensated for just before each injection, when the air is known to be free from H_2S. A Na_2S 0.01 M solution is left in equilibrium with air inside an auxiliary bowl, and part of the air containing the Hs vapors is taken by using a syringe and injected into the reaction chamber to obtain peak concentrations of less than 0.5 ppm. Figure 6 shows on the right an example of traces recorded by the POF sensor and by the electrochemical sensor. On the bottom of Figure 6 is the exposure, i.e., the integral of the electrochemical sensor output numerically computed starting from its values. In the example, it is clear how the POF sensor of this test saturates after the first injection, i.e., with an exposure of less than 2 ppm · hour.

Figure 7 shows the response of a typical POF sensor during a test performed over a long time interval and with H_2S peak concentrations ten-times lower than the ones used for obtaining the traces of Figure 6. The transmittance decreases with an exponential slope as expected from the reaction kinetics, and the sensor reduces its output of about 25% after an exposure to less than 0.1 ppm · hour. In the described test, the average H_2S concentration is about 0.4 ppb. The same figure, on the right, shows one of the arrangements deployed for the sensor, which couples the POF assembly with a battery-operated wireless transmitter and that can be installed inside a showcase without the necessity of any intervention on its topology.

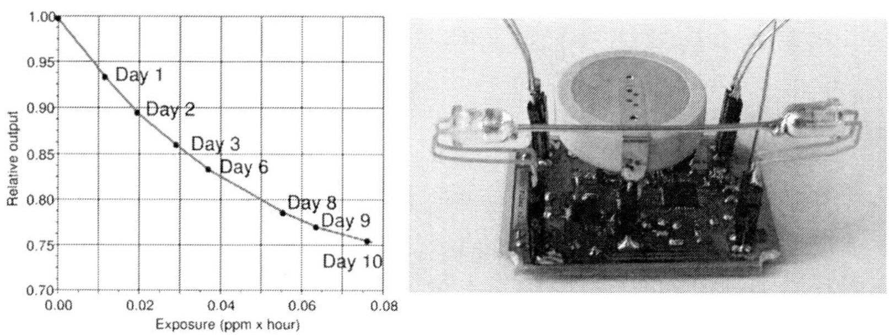

Figure 7. (**Left**) The sensor response to H_2S exposure obtained according to the procedure previously described: each dot represents the transmittance value recorded at the end of each day after an H_2S gas injection; (**Right**) An example of the setup composed of the fiber-based sensor and of the wireless communication system.

4. HF SENSOR OBTAINED VIA PECVD DEPOSITION

This second example describes the deployment of a sensor that exploits an increasing transmittance when exposed to the quantity to be detected. The capability of the fiber to transmit the light is related to differences of refractive indexes between the core and the material deposited onto the core. If a material with a higher refractive index is deposited, the fiber loses part of its capability to transmit the light, and any degradation of this high refractive index coating results in an increase of the transmitted light. This approach has been followed to arrange sensors with the capability of detecting low concentrations of HF vapors.

The sensors described in this section [18,19] have been designed to monitor the concentration of hydrogen fluoride vapors inside the resistive plate chamber (RPC) detectors of the Compact Muon Solenoid experiment at CERN laboratory in Geneva (Switzerland) [20]. These RPCs are composed of two Bakelite plates, maintained at a distance of 2 mm. An electric field of about 4.5 kV/mm is maintained between the plates and an insulating gas mixture composed of $C_2H_2F_4$, C_4H_{10} and % SF_6 flows between the plates. The ionization of this gas upon the passage of a charged particle is the key element of these detectors.

Unfortunately, during the RPC use, the gas mixture gets contaminated due to the production of small amounts of HF, which increases the quiescent RPC current and reduces their detection capability [21]. For this reason, the gas mixture needs to be continuously filtered to remove the HF. The monitoring of the amount of HF to which the gas filters have been exposed is important to avoid filter degradation and to schedule their maintenance.

The material that is unaffected by most compounds and that gets corroded almost only by the HF is the glass. For this reason, to detect the fluoride ions originating from the dissociation of the gaseous HF, the authors decided to employ a glass-like thin film deposited onto the fiber core. SiO_2-like films have a refractive index higher than PMMA; thus, a film of this type on the core makes the fiber quite lossy. In addition, glass-like films are very stable and almost unaffected by quantities different from HF, so that the sensor is extremely selective. Eventually, the reaction between glass and fluoride ions, which leads to the formation of silicon tetrafluoride (SiF_4):

$$SiO_2 + 4HF \Rightarrow SiF_4 + 2H_2O \quad (3)$$

is not reversible; thus, the sensor is inherently suitable for measuring the total exposure to fluoride ions and to estimate the gas filter intoxication without the necessity of numerical post processing.

A stoichiometric SiO_2 glass layer cannot be easily obtained at low temperatures; however, SiO_x thin films can easily be obtained by plasma-enhanced chemical vapor deposition (PECVD) starting from organosilicon monomers. Figure 8 shows a picture of the reactor used by the author for the deposition along with an image of the plasma glow discharge and of the fiber during the deposition.

Figure 8. (**Left**) A picture of the PECVD reactor used for the deposition of the SiO_2-like film; (**Right**) An image of the plasma discharge and of the fiber during the deposition.

The deposition was obtained by using a discharge gas composed of tetraethoxysilane (TEOS) along with O_2 and Ar in different ratios, at a pressure of 5 Pa and with an input power of 50 W. The amount of O_2 in the gas feeding the plasma reactor chamber plays a critical role in the deposition, since a plasma rich in oxygen produces films that are almost inorganic, i.e., with a composition close to SiO_2, while a reduced amount of oxygen leads to organic SiO_x films. The former have a higher refractive index, but are more fragile, while the latter have a lower refractive index, but are much more conformable.

The film thickness can be controlled by varying the deposition time. The maximum sensor sensitivity is obtained with a film thick enough to make the fiber lossy, but thin enough to be quickly attacked by the fluoride ions. After some tests, a value of about 200 nm, which was obtained by using depositions times of about 15 min, was found to be a satisfactory compromise. The thickness of the film was measured by means of an FESEM.

Two different gas compositions, one without oxygen (i.e., with the composition TEOS : O_2 : Ar = 1 : 0 : 30) and one rich in oxygen (i.e., with the composition TEOS : O_2 : Ar = 1 : 20 : 10), were used to obtain two different film types and test for the effect of film composition.

The sensor characterization requires exposing the fiber to an atmosphere containing a known amount of hydrogen fluoride. Even though it is possible to find gas cylinders containing pure hydrogen fluoride gas, the mixing of a known amount of gas with air in order to obtain vary low HF concentrations is quite difficult, so that the author employed the approach that is depicted in Figure 9.

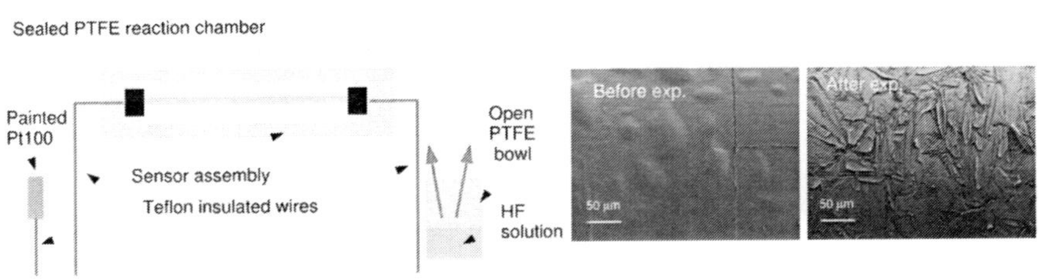

Figure 9. (**Left**) Setup used for sensor characterization; (**Right**) FESEM images of the fiber surface aspect before and after exposure to the HF vapors.

The fiber sensor, along with a temperature sensor, was inserted into a sealed Polytetrafluoroethylene (PTFE) bowl, where also a small bowl containing an aqueous solution of HF acid was inserted. The HF liquid solution evaporates and comes into equilibrium with the air of the sealed reaction chamber. The HF vapor concentration, which is a function of temperature and HF concentration in the solution, can be computed by means of the well-known Antoine law [22]:

$$log_{10}Pp = A - \frac{B}{C+T} \quad (4)$$

where A,B,C are constants connected with the concentration of the HF in the solution, T is the absolute temperature and Pp is the partial pressure of the resulting vapor. The values of the three constants can be found in several places, such as at NIST [23], and can be computed from partial pressure experimental measurements [24].

In the right part of Figure 9 are two FESEM images of a fiber surface before and after exposure to the HF vapors. After deposition and before the exposure, the surface appears smooth and covered by a thin continuous layer of a SiO_2-like film. After the exposure, the surface becomes corroded as SiO_2 is transformed into SiF_4.

Figure 10 shows the fiber transmittance change due to the exposure to HF vapors. The left part of the figure shows two traces referring to two fibers obtained by using the two different discharge gas compositions. Both compositions lead to a thin coating, which is attacked by the fluoride ions, but the layer obtained in the presence of oxygen, and thus much less organic, results in an increased sensitivity that changes from less than 0.01%/(ppm · hour), in the case of the sensor obtained without oxygen, to about 0.02%/(ppm · hour) for the sensor obtained with oxygen presence, thus doubling the sensor sensitivity.

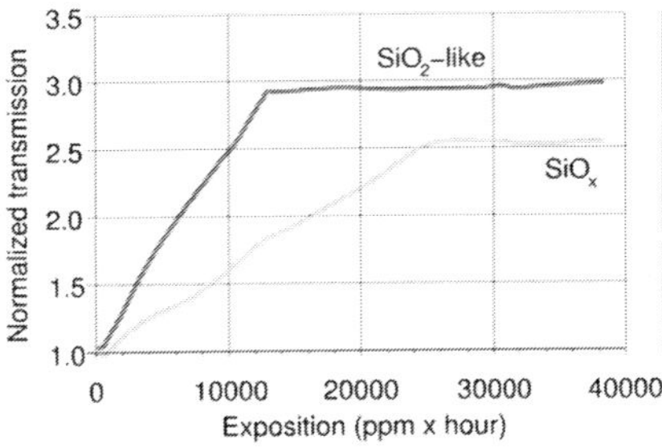

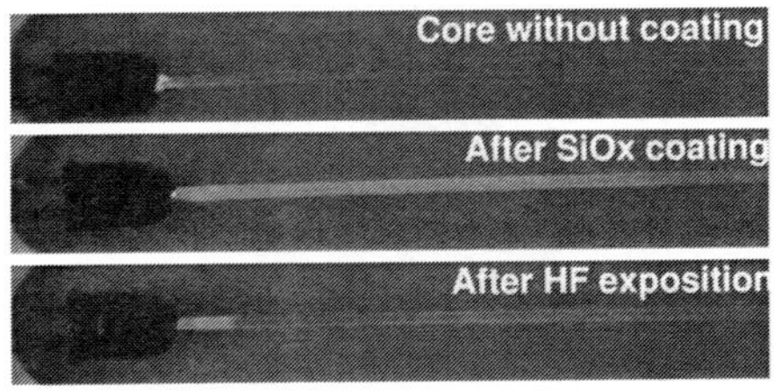

Figure 10. (**Left**) Fiber transmittance change with exposure to HF vapors. (**Right**) Example pictures of the same fiber in three different stages: (**top**) the fiber after cladding removal, but before the SiO_x deposition; (**middle**) the fiber after the SiO_x deposition; (**bottom**) the fiber after exposure to the HF vapors.

In the right part of Figure 10, three pictures of a fiber lit by a red LED in a dark field are reported. The three pictures refer to the same fiber in three different stages: At the top, the fiber after cladding removal, but before the SiO_x deposition. The light is almost completely confined inside the fiber and transmitted to the end. In the middle, the fiber after the SiO_x deposition. A remarkable part of the light is no longer guided by the fiber, so that only a minimal part arrives at the end. At the bottom, the fiber after exposure to the HF vapors. The SiO_x coating is almost completely destroyed, so that the fiber regains most of its capability to transmit the light.

5. CONCLUSIONS

Plastic optical fibers can be used as the key element for developing very low-cost sensors for environmental monitoring by removing the fiber cladding and by coating the core with a thin layer of a substance capable of reacting with the substance to be detected.

Sensors based on the measurement of the fiber transmittance change, even though less accurate than solutions based, for example, on fiber Bragg grating, are extremely inexpensive, have very low energy consumption and, in addition, do not require bulky interrogation systems. These advantages make them specifically suited for installation in adverse conditions and when cabling is not conceivable.

In addition to conventional sensors, which are sensitive to the actual value of the quantity of interest, the POF sensor can be easily designed to be sensitive to the total exposure of a quantity, i.e., to the integral over time of that quantity.

Designing sensors with this behavior requires selecting a coating that exploits a non-reversible chemical reaction with the quantity of interest, but that provides interesting advantages over other conventional techniques based on post processing of the sensor outputs.

In fact, since the cumulative effect is obtained at the physical level, it is possible to avoid all of the problems connected with the presence of offsets, which would greatly impair an integration procedure obtained by post processing the sensor output, and permits detecting and measuring extremely low concentrations of the quantity of interest.

As an example, tests performed on a silver-coated fiber for the detection of H_2S vapors in the cultural heritage field showed the possibility of arranging sensors with a sulfide sensitivity better than 0.2 ppb, i.e., with the capability of monitoring the sulfide concentration in clean air. The described sensor showed an output change of about 0.3%/(ppb · hour), with a useful measuring range of 100 ppb · hour. This means that it is possible to arrange sensors, whose cost is of the order of $1, that can be used to measure the exposure to sulfide vapors in clean air for about one month.

Other tests performed on glass-like coated fibers for the detection of HF vapors for high energy applications proved the possibility of arranging sensors with an output sensitivity of 0.02%/(ppm · hour) and a measuring range of 15, 000 ppm · hour.

The good performance and high sensitivity of the these sensors encourage the development of sensing devices for other applications simply by changing the nature of the substance deposed on the fiber core.

ACKNOWLEDGMENTS

The authors wishes to thank all of researchers from INFNRome and University La Sapienza, Rome, who cooperated in the Italian National Research Project (PRIN2008) titled "Innovative sensors for gas monitoring" for their collaboration in the development and use of the HF sensors.

AUTHOR CONTRIBUTIONS

Sabrina Grassini and Marco Parvis conceived and designed the sensors; Sabrina Grassini performed the depositions and the FESEM analyses; Maen Ishtaiwi performed the experiments; all the authors analyzed the results and contributed to writing the paper.

CONFLICTS OF INTEREST

The author declares no conflict of interest.

REFERENCES

1. Wolfbeis, O.S. Fiber-Optic Chemical Sensors; CRC Press: Boca Raton, FL, USA, 1991; Volume 1–2.

2. Wolfbeis, O.S.; Posch, H.E. Fiber optic fluorescing sensor for ammonia. Anal. Chim. Acta 1986, 11, 185–321.

3. Tubb, A.J.C.; Payne, F.P.; Millington, R.B.; Lowe, C.R. Single-ModeOptical Fiber Surface Plasma Wave Chemical Sensor. Sens. Actuators B Chem. 1997, 41, 71–79.

4. Rhines, T.D.; Arnold, M.A. Fiber-Optic Biosensor for Urea Based on Sensing of Ammonia Gas. Anal. Chim. Acta 1989, 227, 387–396.

5. Schwotzer, G.; Latka, I.; Lehmann, H.; Willsch, R. Optical sensing of hydrocarbons in air or in water using UV absorption in the evanescent field of fibers. Sens. Actuators B Chem. 1997, 38, 150–153.

6. Dietrich, A.M. Measurement of pollutants: Chemical species. Water Environ. Res. 1996, 68, 391–406.

7. Bilro, L.; Alberto, N.; Pinto, J.L.; Nogueira, R. Optical Sensors Based on Plastic Fibers. Sensors 2012, 12, 12184–12207.

8. Stefani, A.; Wu, Y.; Markos, C.; Bang, O. Narrow Bandwidth 850-nm Fiber Bragg Gratings in Few-Mode Polymer Optical Fibers. IEEE Photonics Technol. Lett. 2011, 23, 660–662.

9. Thompson, R.B. Fluorescence-Based Fiber-Optic Sensors. In Topics in Fluorescence Spectroscopy; Lakowicz, J.R., Ed.; Springer: New York, NY, USA, 2002; Volume 2, pp. 345–365.

10. Cennamo, N.; Massarotti, D.; Galatus, R.; Conte, L.; Zeni, L. Performance Comparison of Two Sensors Based on Surface Plasmon Resonance in a Plastic Optical Fiber. Sensors 2013, 13, 721–735.

11. Linec, M.; Donlagic, D. A Plastic Optical Fiber Microbend Sensor Used as a Low-Cost Anti-Squeeze Detector. IEEE Sens. J. 2007, 7, 1262–1267.

12. Merchant, D.F.; Scully, P.J.; Schmitt, N.F. Chemical tapering of polymer optical fibre. Sens. Actuators A Phys. 1999, 76, 365–371.

13. Rovati, L.; Pilati, F.; Ferrari, L.; Fabbri, P. Plastic Optical Fiber pH Sensor Using a Sol-Gel Sensing Matrix; INTECH Open Access Publisher: Rijeka, Croatia, 2012.

14. Corbellini, S.; Parvis, M.; Grassini, S.; Benussi, L.; Bianco, S.; Colafranceschi, S.; Piccolo, D. Modified POF Sensor for Gaseous Hydrogen Fluoride Monitoring in the Presence of Ionizing Radiations. IEEE Trans. Instrum. Meas. 2012, 61, 1201–1208.

15. Arrue, J.; Zubia, J.; Fuster, G.; Kalymnios, D. Light power behaviour when bending plastic optical fibres. IEE Proc. Optoelectron. 1998, 145, 313–318.

16. Vallan, A.; Carullo, A.; Casalicchio, M.L.; Perrone, G. Static Characterization of Curvature Sensors Based on Plastic Optical Fibers. IEEE Trans. Instrum. Meas. 2014, 63, 1293–1300.

17. Angelini, E.; Grassini, S.; Mombello, D.; Neri, A.; Parvis, M.; Perrone, G. Plasma modified POF sensor for in situ environmental monitoring of museum indoor environments. Appl. Phys. A Mater. Sci. Proc. 2012, 100, 975–980.

18. Corbellini, S.; Parvis, M.; Grassini, S.; Benussi, L.; Bianco, S.; Colafranceschi, S.; Piccolo, D. Modified POF Sensor for Gaseous Hydrogen Fluoride Monitoring in the Presence of Ionizing Radiations. IEEE Trans. Instrum. Meas. 2012, 61, 1201–1208.

19. Grassini, S.; Ishtaiwi, M.; Parvis, M.; Benussi, L.; Bianco, S.; Colafranceschi, S.; Piccolo, D. Gas monitoring in RPC by means of non-invasive plasma coated POF sensors. J. Instrum. 2012, 7, 1–11.

20. CMS Collaboration Home Page. Available online: http://cms.web.cern.ch/ (accessed on 27 December 2014).

21. Benussi, L.; Bianco, S.; Colafranceschi, S.; Fabbri, F.L.; Felli, F.; Ferrini, M.; Giardoni, M.; Greci, T.; Paolozzi, A.; Passamonti, L.; et al. Study of gas purifiers for the CMS RPC detector. Nucl. Instrum. Meth. 2011, 661, 241, arXiv:1012.5511.

22. Antoine, C. Tensions des vapeurs; nouvelle relation entre les tensions et les tempè. Comptes Rendus Séances l'Académie Sci, 1888, 107. Available online: http://visualiseur.bnf.fr/ Cadres Fenetre?O=NUMM-3063&=tdm (accessed on 28 December 2014). (In French).

23. NIST Chemistry Web Book. Available online: http:// webbook.nist.gov/chemistry/ (accessed on 28 December 2014).

24. Hydrofluoric Acid Properties. Available online: http://honeywell. com (accessed on 28 December 2014).

CHAPTER 8

Tapered Optical Fiber Sensor for Label-Free Detection of Biomolecules

Ye Tian [1,*], Wenhui Wang [1], Nan Wu [1], Xiaotian Zou [2] and Xingwei Wang [1,2,*]

[1] Department of Electrical and Computer Engineering, University of Massachusetts Lowell, One University Ave., Lowell, MA 01854, USA

[2] Biomedical Engineering and Biotechnology Doctoral Program, University of Massachusetts Lowell, One University Ave., Lowell, MA 01854, USA

ABSTRACT

This paper presents a fast, highly sensitive and low-cost tapered optical fiber biosensor that enables the label-free detection of biomolecules. The sensor takes advantage of the interference effect between the fiber's first two propagation modes along the taper waist region. The biomolecules bonded on the taper surface were determined by demodulating the transmission spectrum phase shift. Because of the sharp spectrum fringe signals, as well as a relatively long biomolecule testing region, the sensor displayed a fast response and was highly sensitive. To better understand the influence of various biomolecules on the sensor, a numerical simulation that varied biolayer parameters such as thickness and refractive index was performed. The results showed that the spectrum fringe shift was obvious to be measured even when the biolayer was only nanometers thick. A microchannel chip was designed and fabricated for the protection of the sensor and biotesting. Microelectromechanical systems (MEMS) fabrication techniques were used to precisely control the profile and depth of the microchannel on the silicon chip with an accuracy of 2 μm. A tapered optical fiber biosensor was fabricated and evaluated with an Immune globulin G (IgG) antibody-antigen pair.

KEYWORDS

optical fiber sensor; biosensor; label-free detection; evanescent wave; MEMS

1. INTRODUCTION

Great efforts have been exerted in the development of optical fiber biosensors for the determination of various analytes, including DNA, proteins, antigens and cells. Those sensors are based on absorbance [1], reflectance [2], fluorescence [3], chemiluminescence [4], bioluminescence [5], or refractive index (RI) measurements [6]. Labels are necessary in most of them. However, the labels cause challenges in biomedical research due to the difficulty of finding an appropriate label that functions equivalently for all molecules [7], the negative effects on the biomolecules, the time-consuming pretreatment and high operation costs. Therefore, for many applications label-free methods are more attractive. Though some label-free biosensors based on surface plasmon resonance (SPR) [8], Fiber Bragg Gratings (FBG) [9], and various interferometers have been developed [10] and exhibit good performance, we intend to introduce a new choice featuring more advantages and potential, which is a simple tapered single-mode fiber biosensor. Its miniature size and mechanical flexibility allow convenient integration with current medical tools for sensing in inaccessible locations when using optical intensity demodulation, and multiple parameters of the tapered fiber profile can be optimized for different applications.

A tapered optical fiber sensor (TOFS) is an excellent candidate for sensing. Single-mode biconical tapered fibers can be applied as physical sensors for measurements of temperature, displacement, and refractive index [11]. They can also be used to couple light into a micro-cavity sensor [12], to be further fabricated as a nanoresonator sensor [13]. A fiber Mach–Zehnder interferometer realized by concatenating two single-mode tapered fibers can also be used as a refractive index sensor [14]. Due to their small size, the aforementioned tapered fibers are attractive for a variety of applications. However, their applications as biosensors are still being developed.

To diversify the applications, more complex tapered fiber structures can be developed. A refractometer can be realized by writing a single long-period grating (LPG) into a biconical tapered fiber [15] or a pair of LPGs into the two sides of the taper region [16]. Some FBG based tapered fibers can measure fluid levels [17]. However, the complex tapered fibers with LPG or FBG still suffer from high fabrication costs. TOFS also have been used for biosensing applications such as the detection of pathogens, protein or cell concentration, and DNA hybridization. Thus far most of TOFS utilize fluorescent labels [18] or SPR [19] to amplify the signal associated with the presence of the bio-target. However, the disadvantages of either a labeling step or additional complexity are undesirable. In this paper, we designed and fabricated a simple tapered single-mode fiber structure to realize a low cost optical fiber biosensor featuring a miniature sensing probe, label-free direct detection, and high sensitivity.

2. PRINCIPLE

Optical fiber usually consists of a core and a cladding with different refractive indices. When light propagates, the main guided field is defined in the core, and the evanescent field is exponentially decayed in the cladding. After a distance of extension, the evanescent field decays to 1/e of its value at the core-cladding interface. As shown in Figure 1, this distance is defined as penetration depth, and it is mathematically described as:

$$d_p = \frac{\lambda}{2\pi\sqrt{n_{co}^2 \sin^2\theta - n_{cl}^2}} \tag{1}$$

where λ is the wavelength of the light source, θ is the angle of incidence of the light at the core/cladding interface, n_{co} and n_{cl} are the refractive indices of the core and the cladding, respectively.

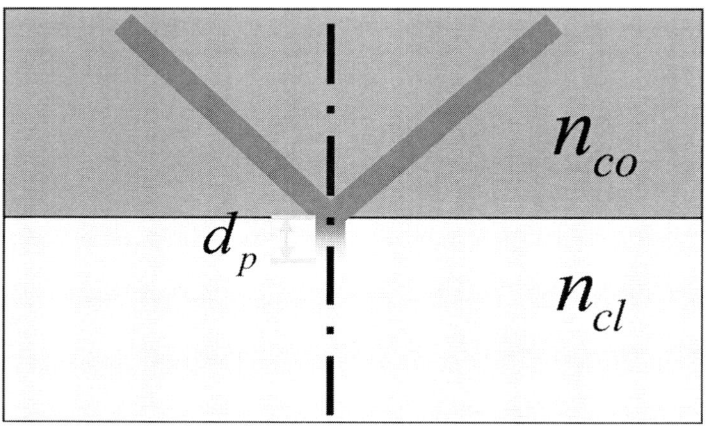

Figure 1. Penetration depth. The evanescent field decays to 1/e of its value at the core-cladding interface.

Optical fibers were originally designed for low loss communication, so the penetration depth is far smaller than the cladding thickness and there is almost no interaction between the optical field and the surrounding environment. For some sensing applications, the evanescent field needs to be exposed to the environment. Tapering fiber is a good solution to make the interaction between the evanescent wave and the surrounding target possible.

Tapered fiber is normally made by pulling the optical fiber when it is heated to its softening temperature to reduce the diameter to tens of micrometers. As a result, the tapered fiber consists of three contiguous parts: one taper waist segment with small and uniform diameter, and two conical transition regions with gradually changed diameter. The ends of conical transition regions are untapered fibers. A typical tapered fiber is shown in Figure 2. In reality, the fiber core material in the waist and transition regions do not have clear boundaries with the cladding material due to their mixing after heating.

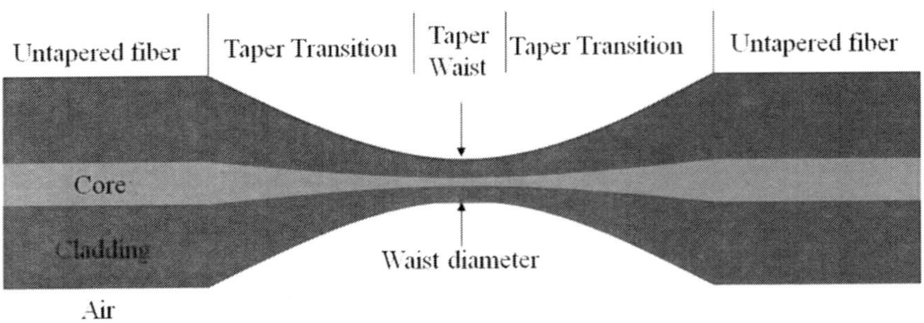

Figure 2. Schematic of a tapered fiber. The tapered fiber consists of three contiguous parts: one taper waist segment with small and uniform diameter, and two conical transition regions with gradually changed diameter. The ends of conical transition regions are untapered fibers.

In the untapered single mode fiber, the thin core is surrounded by the cladding with a lower refractive index. All the light is guided within the core due to total internal reflection. Light propagation through an optical fiber can be described by wave theory. The properties of light in the fiber core are determined by the number of modes which are directly related to V number, given as:

$$V = \frac{2\pi}{\lambda} r(n_{co}^2 - n_{cl}^2)^{1/2} \tag{2}$$

where r is the radius of the core. The only transverse mode of light in the core of untapered single mode fiber is the fundamental mode HE_{11}.

However, at the transition region, along with the decreased diameter, the core of the fiber almost gets mixed together with the surrounding cladding to form a medium whose refractive index is very close to that of the cladding. This medium can be taken as an air-cladding core, which has a larger radius than that of the original single mode fiber core in the most part of the region, and a larger numerical aperture due to larger refractive index difference between the cladding and the air. According to Equation (2), this region functions like a multimode fiber that supports multiple modes. This transition region is normally divided into two distinct categories: adiabatic and nonadiabatic. In the condition of adiabatic transition, the fundamental fiber mode HE_{11} can be carried out with efficiency as high as 99.5% and the contribution of higher order modes is insignificant and is not taken into consideration [19]. In the condition of nonadiabatic transition, some high-order modes can be excited, but the first two modes HE_{11} and HE_{12} are significant if the V number is controlled properly.

In the taper waist region, part of the light energy is not confined by the thin waist, around which an evanescent field is generated, and primarily the first two modes HE_{11} and HE_{12} coupled from one end of nonadiabatic transition region propagate at the air-cladding interface.

After the interference, HE_{11} and HE_{12} modes couple back into fundamental mode at the other end of nonadiabatic transition region. The second coupling efficiency depends on relative phase $\Delta\emptyset = \Delta k \cdot l$, where Δk is relative propagation constant and l is coupling distance. Δk can be tuned by changing the environment medium. l can be changed by taper waist length. The interference effect between HE_{11} and HE_{12} along the waist region incurs fringes in transmission spectrum, typically a sinusoidal behavior [20]. If Δk or l changes, the spectral response of the taper will shift correspondingly [11]. Such interference and spectral response shifts are used for the biosensing in this paper. In order to ensure this interference, the taper waist diameter needs good control. If it is less than one micrometer approximately the interference effect does not exist for 1,550 nm light, because at such size the waist has similar properties as the core in the single mode fiber and it no longer supports HE_{12} mode. This diameter limit can be calculated by substituting the single mode V number limit (2.405), and the refractive indices of cladding and air into Equation (2). Therefore, we normally control the taper waist to be about ten micrometers in diameter.

3. FABRICATION

3.1. Tapered Sensor Fabrication

The parameters such as taper waist diameter, length, and transition shape, are critical for a specially required interference spectrum. By controlling the fabrication conditions such as pulling speed, heating length and temperature, tapered fiber with different shapes and properties can be fabricated. Figure 3 becomes a typical fabrication setup. Single-mode fibers (SMF) with core/cladding diameters of 8/125 μm were used. Its protective coating was removed over a section of about 15 mm long for heating. Two clamps were used to fix the SMF onto two translation stages (NRT 100, Thorlabs). A torch flame or CO_2 laser was used as a heating source. A laser in a component test system

(CTS) (Si 720, Micron Optics) excited one end of the fiber. The spectrum response from the other side of the fiber was monitored.

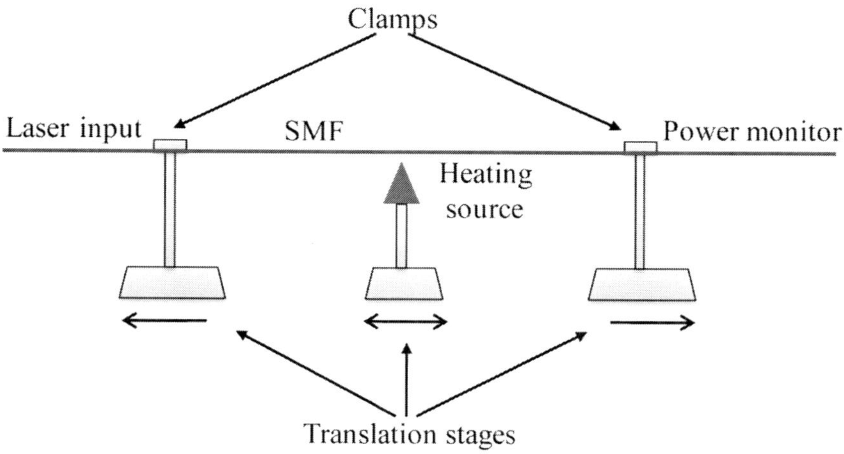

Figure 3. A typical fabrication setup. Two clamps were used to fix the SMF onto two translation stages. A torch flame or a CO_2 laser could be utilized as heating source. A CTS excited one end of the fiber and monitored the spectrum response from the other side.

The fiber could be pulled from one end [11] or both ends [19], and the heating source could be fixed [14] or oscillated [16]. In order to prevent too many variable factors that may incur uncertainties in the optical performance of the tapered fiber, we only pulled one end. For the heating source, because a torch flame could not provide repeatable heating size and temperature, an Engraving Systems (Zing-Model 10000) CO_2 laser was applied. Its laser source was digitally controlled and permanently aligned. Its adjustable laser power, speed, and pulse frequency provided much heating flexibility. In this case, we applied an oscillation heating with frequency 2 Hz within a 5 mm fiber section.

During the fabrication process, the fiber must be held straight all the time. After one second of heating startup, the translation stage was started with acceleration of 0.3 mm/s^2 and constant speed of 0.9 mm/s to move a defined length. Transmission spectrum data was collected by a spectrum response recording system for further analysis. Figure 4 shows a typical fabricated fiber.

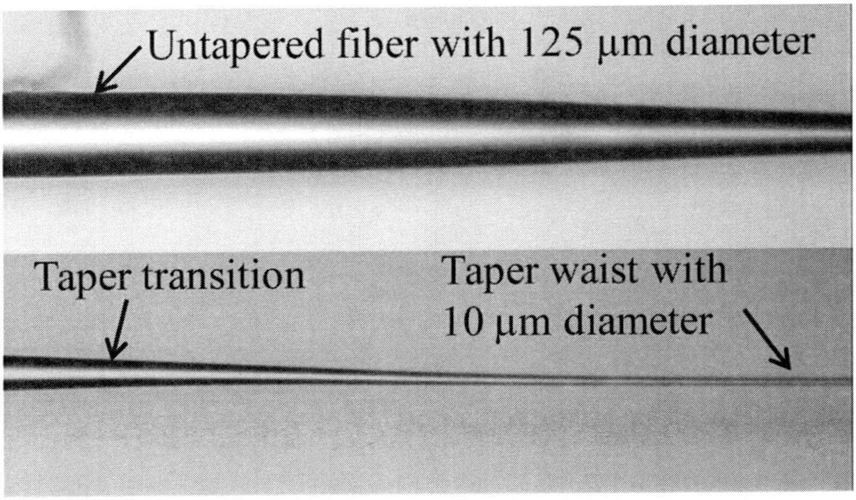

Figure 4. A typical fabricated tapered fiber. The untapered fiber was 125 μm of diameter, and the taper waist was about 10 μm of diameter. The taper transition region had a gradually changed diameter.

Tapered fiber could also be fabricated by a special programmable fusion splicer with discharge electrode [21] or filament. An optical fusion splicer (S177A) was used for fabrication. This method was automatic, but the tapered length was limited, which could reduce the biosensing sensitivity.

3.2. Microchannel Holder Fabrication

The tapered fiber sensor was thin and fragile after the fabrication. A microchannel chip was designed and fabricated for the sensor protection and further biotesting. The schematic of the microchannels on a silicon chip was shown in Figure 5(a).

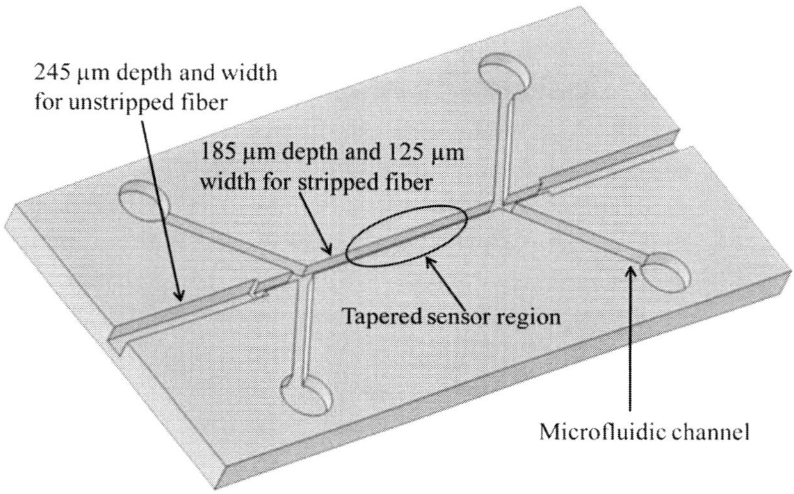

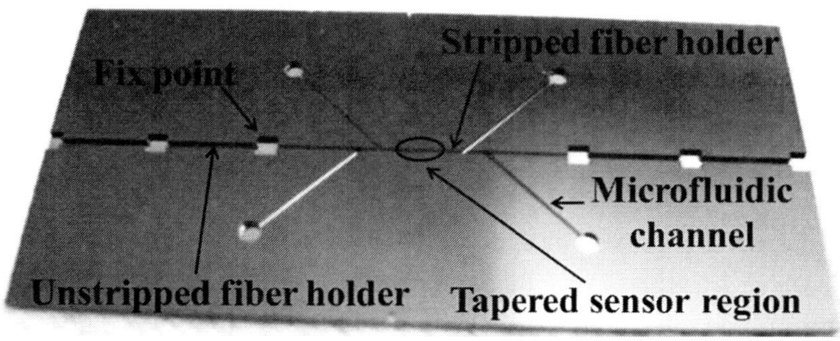

Figure 5. (a) (upper) Chip structure. Both ends of the channel were in 245 μm depth and width so they could hold the unstripped sections of the fiber, while the middle of the channel was in 185 μm depth and 125 μm width to hold the stripped fiber. The four microfluidic channels branches were designed for inlets and outlets of bio-target and fluid, respectively. **(b)** (lower) Real etched chip.

Both ends of the channel were in 245 μm depth and width so they could hold the unstripped sections of the fiber, while the middle of the channel was in 185 μm depth and 125 μm width to hold the stripped fiber. The four microfluidic channels branches were designed as inlets and outlets for the bio-target and fluid, respectively.

Microelectromechanical systems (MEMS) fabrication techniques, specifically photolithography and deep dry etching, were used to precisely control the profile and depth of the microchannel on the silicon chip with an accuracy of 2 μm. First, a mask with the desired profile of microchannels and microfluidic system was fabricated, and the mask pattern was transferred to a silicon wafer by a layer of photoresist. Then, the patterned wafer was etched with an inductively coupled plasma deep etcher. Figure 5(b) showed a real etched chip. The channel sizes were matched to fit the fiber dimensions. After the sensor was fixed into the microchannel chip, the chip surface was sealed by a layer of polydimethylsiloxane (PDMS) leaving the inlets and outlets open.

4. RESULTS AND DISCUSSION

4.1. Experiments

The biotest was set up as shown in Figure 6. The component test system was used to excite laser into one end of the tapered fiber sensor, and monitor the transmission spectrum from the other end. The bio-target bonding in the tapered region could be analyzed through the phase shifts of the spectrum.

For a typical test, IgG antibody-antigen pair was used to evaluate the performance of the tapered fiber biosensor. The biotest procedure was as follows: the tapered fiber was washed with deionized water and dried in air or by nitrogen gas; The 2% aminopropyltriethoxysilane solution in acetone was prepared for silanization; the prepared solution was injected into the sensing region for 30 minutes at room temperature; the resulting alkylaminosilane-derivatized tapered surface was cleaned by deionized water; a solution of 10% glutaraldehyde in 50 mM sodium phosphate-buffered saline (PBS) at pH 7.4 was injected onto the alkylaminosilane-derivatized surface for 2 hours to activate the silanized substrate, followed by washing out the excess of glutaraldehyde thoroughly with deionized water and drying; IgG antibody solution was injected onto the activated surface and kept for 2 hours at 4 °C, followed by washing out the physically adsorbed IgG antibody with deionized water and drying; The initial transmission spectrum of antibody layer was recorded by the CTS system; IgG anitigen solution was injected to the oriented antibody layer and kept for 2 hours at 4 °C, followed by washing out the remaining nonimmobilized antigen with deionized water and drying; The transmission spectrum of IgG antibody-antigen pair layer was recorded for comparison.

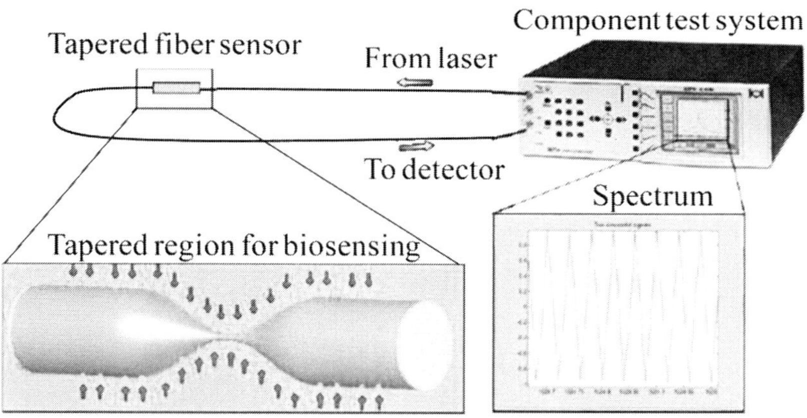

Figure 6. The CTS was used to excite laser into one end of the tapered fiber sensor, and monitor the transmission spectrum from the other end. The bio-target bonding in the tapered region could be analyzed through phase shift of the spectrum.

The comparison between the initial transmission spectrum (black) and IgG antibody-antigen pair layer (red) was shown in Figure 7(a). There was an obvious redshift of 1.50 nm after the capture of the IgG antigen. The reason of the small changes of profile and intensity in spectrum response will be analyzed after a comparison with the simulation results. As the repeatability test results shown in Figure 7(b), the captures of IgG antigen were successfully detected by demodulating the peak wavelength redshift of 1.50 nm and 1.47 nm, respectively, in two trials.

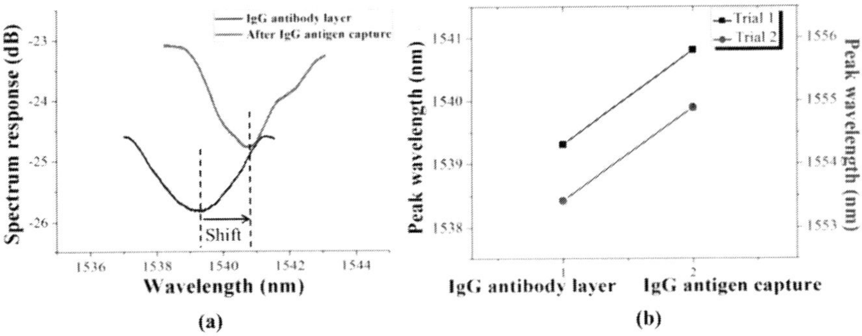

Figure 7. (a) The comparison between the initial transmission spectrum (black) and IgG antibody-antigen pair layer (red). An obvious redshift of 1.50 nm after the capture of the IgG antigen was monitored. **(b)** The captures of IgG antigen were successfully detected by demodulating the peak wavelength redshift of 1.50 nm and 1.47 nm, respectively in two trials.

4.2. Simulations

To better understand the influence of various biomolecules on the sensor, a numerical simulation that varied the thickness and refractive index of the bio-layer was performed. According to the dimensions and properties of the tapered fiber, a three-dimensional (3D) tapered fiber model was built to simulate the transmission spectrum response based on the Beam Propagation Method (BPM).

In the simulation model, taper waist length was 12 mm, transition lengths were 4 mm and 8 mm, and transition profile was exponential or conical. Untapered fiber diameter was 125 μm, and taper waist diameter was 10 μm. Background refractive index n_b = 1, cladding refractive index n_{cl} = 1.4629, core refractive index n_{co} = 1.4682. The transmission spectrum fringe redshift when a layer with refractive index 1.33 was coated on the taper waist surface is shown in Figure 8. The black curve indicated the transmission spectrum under a bio-test preparation layer with refractive index 1.33 and thickness 100 nm. After a target biolayer with refractive index 1.33 and thickness 100 nm was bonded upon the preparation layer, the fringe shifted 3.5 nm to the red curve.

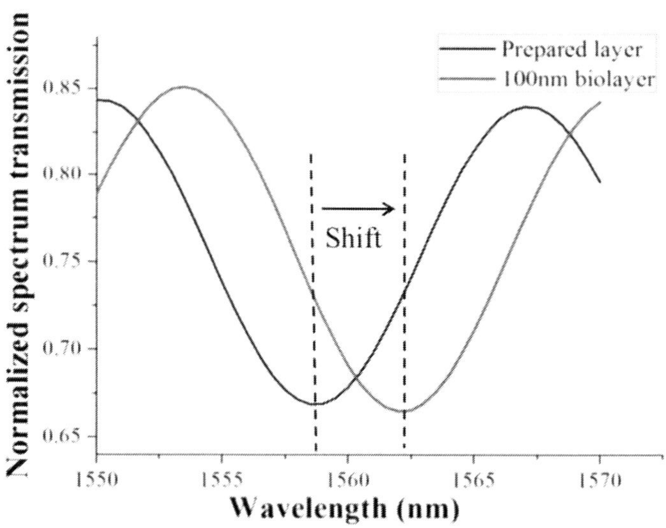

Figure 8. The transmission spectrum fringe redshift. The black curve indicated the transmission spectrum under a bio-test preparation layer with refractive index 1.33 and thickness 100 nm. After a target biolayer with refractive index 1.33 and thickness 100 nm was bonded above the preparation layer, the redshift 3.5 nm was represented by the red curve.

In general, the bonded biolayer should be in the nanometer range. To illustrate the potential to detect a very thin bio-layer, an optimized tapered fiber model with thinner and longer waist was used to calculate the influence of the presence of a thin biolayer at the taper waist. Figure

9 showed the phase shift with different biolayers' thicknesses at the nanometer level. The conditions of three typical refractive indexes were given. Each of the three conditions has a linear wavelength shift response to the bonded biolayer thickness.

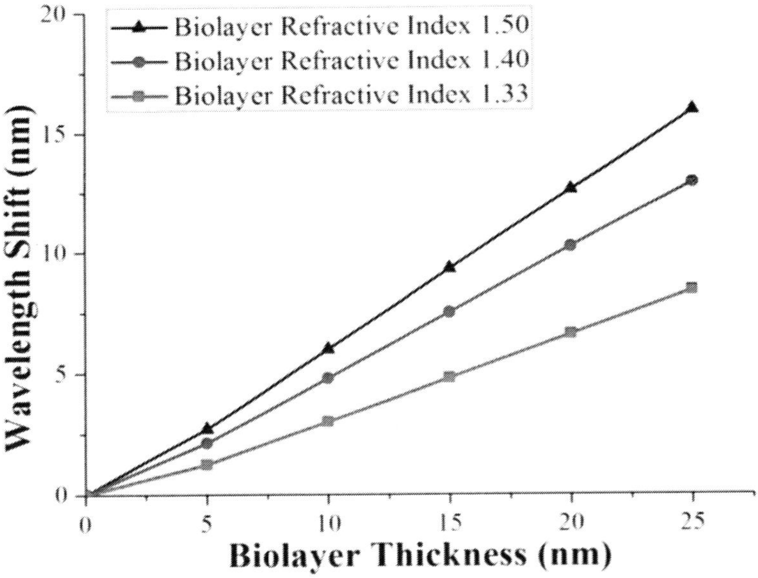

Figure 9. The phase shift with different biolayers thickness in nanometer level. Each of the three conditions has a linear phase shift response to the bonded biolayer thickness.

4.3. Analysis

The computer simulation showed perfect sinusoid spectrum profiles and redshifts after the bonding of biomolecule layers. However, the experimental results showed small changes in profile and intensity besides the obvious redshift. There might be some reasons for this. First, the tested tapered fiber sensor was not as ideal as the simulation model with the uniform taper waist and exponential taper transitions, so the spectrum was not a regular sinusoid due to more high-order propagation modes. Second, the biomolecule layers were not only bonded on the taper waist, but also the taper transitions which might change the propagation coefficient of the waveguide. Third, there might be physically adsorbed antigen which also changed the properties of the tapered waveguide. Fourth, a small intensity change was demonstrated in the simulation results (Figure 8), but the redshifts were close to linear (Figure 9), so the biosensing could be quantified from the linear redshifts, and the small intensity change in the experiment could be ignored. In all, a better experimental result should be obtained by further improved fabrication of the tapered fiber sensor and the more precise control of the biotesting.

5. CONCLUSIONS AND FUTURE WORK

A low cost tapered fiber biosensor featuring a miniature sensing structure, label-free direct detection, and high sensitivity was presented. A miniature biosensor less than ten micrometers wide and several millimeters long was fabricated. Its high sensitivity was realized by the sharp spectrum fringe and sufficiently long biomolecule reaction region. A microchannel chip was designed and fabricated for the sensor's protection and biotesting. Finally, the good performance of the fabricated biosensor was evaluated with an IgG antibody-antigen pair, and the influence of the biolayer on the tapered fiber was simulated to consolidate the theoretical analysis and experimental results. To minimize the whole system and lower the cost, an intensity interrogator is under study.

ACKNOWLEDGMENTS

The authors gratefully appreciate the financial support for this work provided by the Massachusetts Life Sciences Center (00000000000 3044).

REFERENCES

1. Lucas, P; Solis, MA; Coq, DL; Juncker, C; Riley, MR; Collier, J; Boesewetter, DE; Boussard-Pledel, C; Bureau, B. Infrared biosensors using hydrophobic chalcogenide fibers sensitized with live cells. Sens. Actuat. B Chem 2006, 119, 355–362.

2. Watanabe, M; Kajikawa, K. An optical fiber biosensor based on anomalous reflection of gold. Sens. Actuat. B Chem 2003, 89, 126–130.

3. Nguyen-Ngoc, H; Tran-Minh, C. Fluorescent biosensor using whole cells in an inorganic translucent matrix. Anal. Chim. Acta 2007, 583, 161–165.

4. Magrisso, M; Etzion, O; Pilch, G; Novodvoretz, A; Perez-Avraham, G; Schlaeffer, F; Marks, R. Fiber-optic biosensor to assess circulating phagocyte activity by chemiluminescence. Biosens. Bioelect 2006, 21, 1210–1218.

5. Gautier, SM; Blum, LJ; Coulet, PR. Fibre-optic biosensor based on luminescence and immobilized enzymes: Microdetermination of sorbitol, ethanol and oxaloacetate. J. Biolumineśc. Chemiluminesc 1990, 5, 57–63.

6. Tian, Y; Wang, W; Wu, N; Zou, X; Guthy, C; Wang, X. A miniature fiber optic refractive index sensor built in a MEMS-based microchannel. Sensors 2011, 11, 1078–1087.

7. Caruso, F; Rodda, E; Furlong, DN; Niikura, K; Okahata, Y. Quartz crystal microbalance study of DNA immobilization and hybridization for nucleic acid sensor development. Anal. Chem 1997, 69, 2043–2049.

8. Esteban, O; Gonzzlez-Cano, A; Diaz-Herrera, N; Navarrete, MC. Absorption as a selective mechanism in surface plasmon resonance fiber optic sensors. Opt. Lett 2006, 31, 3089–3091.

9. Tang, JL; Cheng, SF; Hsu, WT; Chiang, TY; Chau, LK. Fiber-optic biochemical sensing with a colloidal gold-modified long period fiber grating. Sens. Actuat. B Chem 2006, 119, 105–109.

10. Wang, X; Cooper, KL; Wang, A; Xu, J; Wang, Z; Zhang, Y; Tu, Z. Label-free DNA sequence detection using oligonucleotide functionalized optical fiber. Appl. Phys. Lett 2006, 89, 163901.

11. Kieu, K; Mansuripur, M. Biconical Fiber Taper Sensors. IEEE Photo. Tech. Lett 2006, 18, 2239–2241.

12. Vahala, KJ. Optical microcavities. Nature 2003, 424, 839–846.

13. Tong, L; Gattass, RR; Ashcom, JB; He, S; Lou, J; Shen, M; Maxwell, I; Mazur, E. Subwavelength-diameter silica wires for low-loss optical wave guiding. Nature 2003, 426, 816–819.

14. Tian, Z; Yam, SSH; Barnes, J; Bock, W; Greig, P; Fraser, JM; Loock, HP; Oleschuk, RD. Refractive index sensing with Mach-Zehnder Interferometer based on concatenating two single-mode fiber tapers. IEEE Photo. Tech. Lett 2008, 20, 626–628.

15. Allsop, T; Floreani, F; Jedrzejewski, KP; Marques, PVS; Romero, R; Webb, DJ; Bennion, I. Spectral characteristics of tapered LPG device as a sensing element for refractive index and temperature. J. Lightw. Tech 2006, 24, 870.

16. Ding, JF; Zhang, AP; Shao, LY; Yan, JH; He, S. Fiber-taper seeded long-period grating pair as a highly sensitive refractive-index sensor. IEEE Photo. Tech. Lett 2005, 17, 1247–1249.

17. Grobnic, D; Mihailov, SJ; Huimin, D; Smelser, CW. Bragg grating evanescent field sensor made in biconical tapered fiber with femtosecond IR radiation. IEEE Photo. Tech. Lett 2006, 18, 160–162.

18. Fielding, AJ; Davis, CC. Tapered single-mode optical fiber evanescent coupling. IEEE Photo. Tech. Lett 2002, 14, 53–55.

19. Villatoro, J; Monzon-Hernandez, D; Mejia, E. Fabrication and modeling of uniform-waist single-mode tapered optical fiber sensors. Appl. Opt 2003, 42, 2278–2283.

20. Kieu, K; Mansuripur, M. Tuning of fiber lasers by use of a single-mode biconic fiber taper. Opt. Lett 2006, 31, 2435–2437.

21. Leung, A; Shankar, PM; Mutharasan, R. Label-free detection of DNA hybridization using gold-coated tapered fiber optic biosensors (TFOBS) in a flow cell at 1,310 nm and 1,550 nm. Sens. Actuat. B Chem 2008, 131, 640–645.

CHAPTER 9

Optical Fiber Torsion Sensor with Mechanically Induced Long Period Fiber Gratings in Rare-Earth Doped Fibers

Maria Pulido-Navarro[1*], José Álvarez-Chávez[1], Daniel Ceballos-Herrera[2], Ponciano Escamilla-Ambrosio[3]

[1]Instituto Politécnico Nacional, Centro de Investigación e Innovación Tecnológica (CIITEC-IPN), Mexico City, Mexico

[2]Universidad Autónoma de Nuevo León (UANL), Nuevo León, Mexico

[3]Instituto Politécnico Nacional, Centro de Investigación en Computación (CIC-IPN), Mexico City, Mexico

ABSTRACT

In this work wavelength sensitivity in mechanically induced long period fiber gratings (MLPFG) is analyzed. This analysis is first carried out both in standard single-mode fiber SMF-28 and in Er-doped fibers. The mechanical analysis for both types of fibers under different torsion conditions is presented. In order to apply the torsion one of the fiber ends is fixed while torsion is applied on the other end. A MLPFG whose period is 503 μm is used to press the fiber after torsion is applied. This allows for micro curvatures to be formed on the fiber, which in turn generates a periodical index perturbation on it. Here, it was noted that the sensitive wavelength shift of the rejection bands is bigger for Er-doped fibers. For a torsion of 6 turns applied to 10 cm of doped fiber the wavelength peaks can be moved up to 25 nm, which is longer to what was detected on standard fibers. Therefore, by using Er-doped fibers to monitor torsion on structures will give more sensitive and accurate results than using standard fibers. These results can be employed for sensing applications, especially for small to medium size structures, which can be mechanical, civil or aeronautics.

KEYWORDS

Optical Fiber Sensors, Fiber Bragg Gratingsensors, Torsion Sensors, Rare-Earth Doped Fibers

1. INTRODUCTION

Torsion sensors are useful in the analysis of the working status of structures. Conventional torsion sensors are normally based on angle measurement devices, such as optical encoders and magnetic sensors [1] . These types of sensors have a large size and are usually difficult to be embedded into structures. The concept of smart structure is one in which the structure has been provided with a system that is capable of real time and in-situ measurement. Recently, sensors that can be embedded into structures have drawn a lot of attention because they can provide the technological basis for the development of smart structures [2] . Sensors based on optical fibers can be embedded or attached to structures allowing for higher sensitivity to external changes and thus exhibit better reliability. There are many types of sensors based on optical fibers. They are classified into three main types: interferometric, distributed and sensors based on Bragg gratings. An interferometric sensor consists of an optical cavity which goes from one end of the fiber to the surface to be analyzed. Two beams of light are employed; they travel along two different optical paths, made up of a system of mirrors that finally converge in order to form an interference pattern. The physical changes in structures are reflected by the changes in the optical phase between the interference of two lightwaves [3] . A distributed sensor refers to a kind of sensor where the optical fiber itself is the sensor, with lengths in the range of kilometers [4] . Distributed sensors are based on backscattering. They use this phenomenon as their measurement basis in order to improve sensitivity in the detection of discontinuities in structures. Scattering is divided into three phenomena: Rayleigh backscattering, Brillouin scattering and Raman, either stimulated or spontaneous [5] -[7] . Finally, Bragg gratings are periodic perturbations of the core of the fiber, where the distance between perturbations is called period (Λ). Within sensors based on Bragg gratings there are two types of gratings: short and long period fiber gratings (LPFG). Short period Bragg gratings show periodic perturbations with periodicities under 100 μm [8] . For LPFG the periodicities are from 100 μm up to 1000 μm [9] [10] . In LPFGs when light is incident into the fiber, the light couples into the co-propagating higher order modes producing, in the transmission spectrum, a series of missing bandwidths called resonance wavelengths. The resonance wavelengths depend directly on the Bragg period and the refraction index of the core of the fiber. These parameters are affected by external environmental changes such as temperature, humidity or strain. When this change occurs, the resonance wavelength shifts. By measuring this shift it is possible to relate the change in wavelength with strain, temperature or humidity. One advantage of LPFGs is that it is possible to have hundreds of sensors in the same optical fiber. LPFGs present some advantages over FBG such as simplicity of fabrication, low back reflection, low insertion loss and high sensitivity to temperature, bending and torsion [11] .

2. LONG PERIOD FIBER GRATINGS FABRICATION METHODS

Many methods have been studied and proposed for the fabrication of LPFGs. Ultraviolet (UV) laser exposure, is the preferred technique because of its ease of fabrication and mass production. With this method the amount of index change depends directly on the intensity and duration of the UV laser. The resulting index modulation is symmetrical and as a result of this, only axially symmetric cladding modes are coupled. The greatest disadvantage of this method is that the fiber needs to pass through a process of sensitization before manufacturing the gratings. Furthermore with time they tend to decay and cannot stand temperatures in the order of 600°C [12] . Other proposed methods such as CO_2 laser irradiation, mechanical pressure, etched corrugations, electric-arc discharge, ion beam implantation and femtosecond laser exposure are asymmetric techniques [13] [14] . CO_2 laser irradiation at 10.6 μm produces gratings with high temperature stability. For the etched corrugation a tensile load is applied to the fiber resulting in a periodic strain variation in the fiber which causes a periodic modulation of the refractive index through the photoelastic effect. Electric-arc discharge is a method that uses a commercial fusion splicer, which has as its main advantages its simplicity and low cost. The splicer creates the index modulation on the fiber by local heating. The femtosecond technique uses pulses in the near infrared around 800nm. The irradiation causes densification of the glass and the gratings can stand temperatures up to 500°C [15] . Compared with other methods, mechanical pressure is a flexible technique of low cost and also no previous process to prepare the fiber is required. Furthermore, this technique benefits from being a reversible process [16] , which means that once the mechanical piece is removed, the mechanical integrity of the fiber is kept. This is quite useful for research work and for field devices.

3. THEORY

LPFGs contain a series of attenuation bands centered at discrete wavelengths. The attenuation is due to the coupling of light from the fundamental core mode (LP_{01}) to selected cladding modes. The only cladding modes (LP_{mk} o HE_{1k}) to which the guided core mode can couple are those that exhibit azimuthally symmetric intensity profiles with central peaks, where m = 0 and k = (1, 2, 3,...). Light coupled to the cladding at specific wavelength modes is attenuated or radiated out giving as a result rejections bands in the transmission spectrum [17] . The phase-matching condition for LPFGs is [18] :

$$\beta_{core} - \beta_{cladd}^{m} = \frac{2\pi}{\Lambda} \tag{1}$$

where β_{core} and β_{cladd}^{m} are the propagation constants of the fundamental core mode and the cladding mode, respectively; Λ is the grating period, and m is the order of cladding modes. Therefore, the resonance condition for coupling is given by [17] :

$$\lambda_{res} = \left(n_{core} - n_{clad}^{m}\right)\Lambda \tag{2}$$

where λ_{res} is the initial resonant wavelength, n_{core} and n_{clad}^{m} are effective indexes of the fundamental core mode and cladding mode, respectively.

The minimum transmission of the attenuation band is given by [19] :

$$T_i = 1 - \sin^2\left(\kappa_i L\right) \tag{3}$$

where L is the length of the LPFG and κ_i is the coupling coefficient for the ith cladding mode.

For telecommunications purposes, low sensitivity is desired because it will minimize the influence of environmental conditions. On the contrary, for sensors systems, high sensitivity is seen as an advantage. An important characteristic of LPFGs is their sensitivity, so full characterization is essential. In a natural environment, the LPFG characteristics are affected as the external conditions change. The sensing effects are primarily due to a differential change induced in the two modes. More specifically, since the effective indexes n_{core} and n^{m}_{clad} undergo dissimilar changes owing to a change in the external conditions, the difference between the effective indexes of the two modes Δn_{eff}, is altered. What is more, a fixed periodicity of the LPFG shrinks or stretches resulting from a change in external conditions as well. Therefore, the resonance wavelength of the LPFG has got to shift due to the change of Δn_{eff} and/or Λ. Consequently, the change of external parameters such as temperature, strain, bending, torsion and load, can be determined by measuring or calculating the resonance wavelength shift of the LPFG.

4. EXPERIMENT

In this section the torsion sensitivity of a Mechanical LPFG (MLPFG) on a standard single mode fiber SMF-28 and an Erbium-doped fiber is investigated. MLPFG is made up of two plates; one flat and the other grooved. For this experiment the grooved plate was done on an acrylic slab with the help of a CNC (Computer Numerical Control). In this case the change in the core refractive index of the fiber is accomplished by putting the fiber in between the two plates, and as can be appreciated from Figure 1, the fiber acquires the form of the grooved plate slots. MLPFG present the benefit of tunability; it means that the grating period can be modified by rotating the angle α between the grooved plate and the longitudinal axis of the fiber, as shown from the top view of the mechanical device in Figure 1. For our experiment a mechanical plate of 4cm length with a grooved period Λ of 503 μm is used. The grooved depth is 300 μm and its resonance wavelength and amplitude are 1530 nm and −62 dB, respectively. The length of the tested fiber is about 10 cm. The two ends of the fiber were fixed at two holders; one end of the fiber is twisted by rotating a disc attached to one of the holders by N circles. The fiber is twisted before being pressed with the grooved plate. The twist can be removed by rotating the disc in opposite direction or through releasing the fiber from the holders, which allows the fiber to recover its original mechanical properties.

A white light and a light source at 980 nm in amplified spontaneous emission (ASE) were used for the standard and the doped fibers, respectively. An optical spectrum analyzer (OSA) was used to measure the transmission spectrum as the fibers were twisted. On the other hand, by placing the mechanical induced grating at the right angle α, the peaks of the grating can be tuned until they are observed in the ASE spectrum, as can be seen in Figure 2 and Figure 3. What is important on this instance is to observe how the peaks are formed when the MLPFG is positioned before torsion perturbations are applied. After this step and for comparison purposes, before pressing the fiber with the grooved plate, the fiber is twisted with the rotating holder and positioned over the flat plate as can be seen in Figure 1. Subsequently, the torsion effect is shown on the four peaks of the grating. Before working on the doped fiber, experiments were carried out first on the standard SMF-28 fiber. From Figure 2, it can be observed that after 6 rotations the wavelength shifted around 3 nm, from 1223.8 nm to 1227.4 nm. After that, the laser is operated in ASE for the doped fiber. From Figure 3, four peaks are observed just before rotation is performed (1510, 1522, 1565 y 1640 nm). From Figure 4 and after torsion is applied, it is detected that the peaks shifted. This figure shows both peaks: before and after rotation, the only purpose of getting this figure is for a better analysis of the peaks position. From the results in Figure 4 it can be observed that the four peaks shift to smaller wavelengths and every peak shifts at different rate. In order to observe in a more detailed way this rate changes, the movement of the peaks for different fiber twisting are shown in the graph from Figure 5.

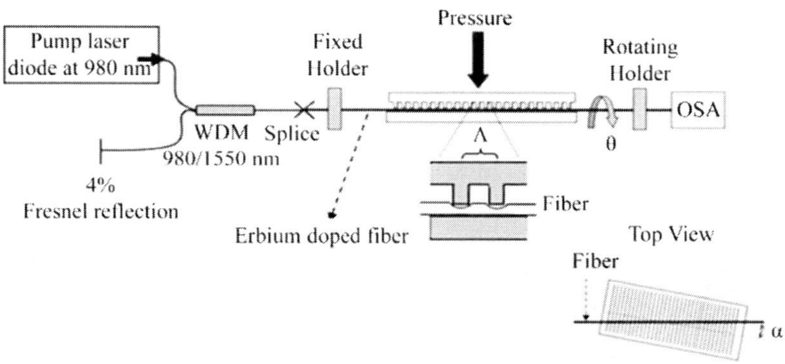

Figure 1. Experimental setup for torsion characterization of mechanically induced LPFG in Er-doped fibers.

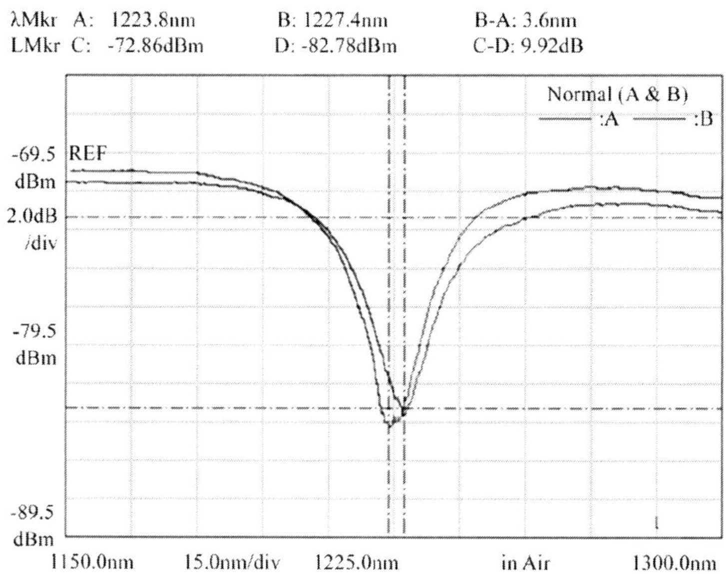

Figure 2. Wavelength shift in standard SMF-28 with MLPFG.

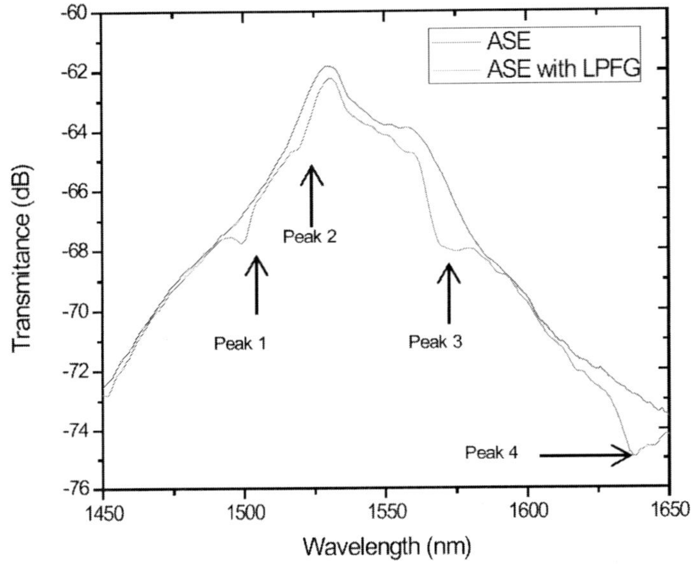

Figure 3. ASE Spectra of the Er-doped fiber with the mechanically induced LPFG.

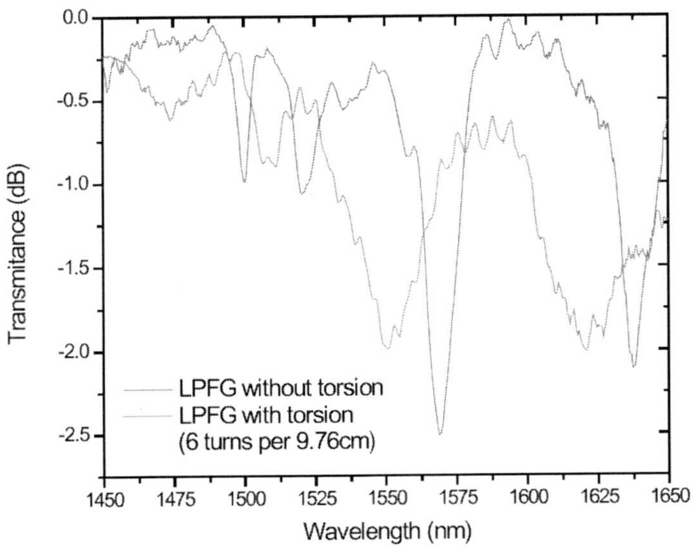

Figure 4. ASE Spectra of the Er-doped fiber with the mechanically induced LPFG under torsion.

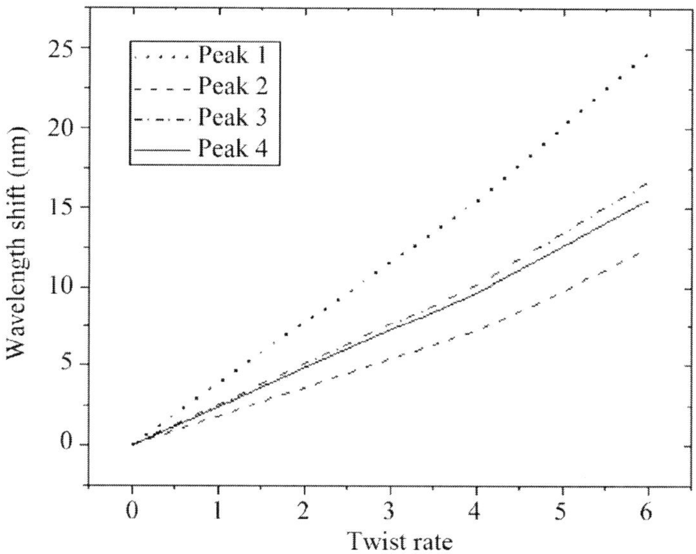

Figure 5. Wavelength shift of each peak of the LPFG at different twist rates.

From Figure 5 it can be observed that the shift on the central wavelength of the peaks with different values of twisting have a linear behavior. All the measurements made in order to obtain the Figure 3 and Figure 4 were done using an OSA with a resolution of 0.2 nm. Also from Figure 5 it can be observed that for a twist of 6 laps done onto a piece of doped fiber of 10 cm length, the peaks shifted up to 25 nm. The aforementioned in particular improves what has been reported for standard fiber SMF-28. Therefore, it can be stated that Erbium doped fibers are superior on sensing twisted strain in structures.

5. CONCLUSION

This work was done with the idea to prove that the sensitivity of LPFGs under twisted circumstances is increased when using doped fibers instead of standard fibers. The aforementioned was corroborated; it was found that the sensitivity is highly increased when using MLPFG on doped fibers and it was observed that the peak wavelengths can shift up to 25 nm. MLPFG are easy to fabricate, cost effective and also tunable and reversible. The pressure applied on the LPFG changes the refractive index of the fiber core, thus, the grating period can be easily tuned by changing the angle α between the fiber and the grooved plate, which in turn will vary the resultant resonance wavelength. In future experiments, the resonance wavelength could be easily controlled via the grating period Λ. Furthermore, the period depth plays a role in the mode coupling efficiency. It was observed that for period depths of 300 μm and above led to stronger mode coupling. It shows deeper dips which is the result of induced higher index variation. For this work, mechanical LPFGs were used for study purposes, once the right grating period is determined, the in-fiber Bragg grating can be achieved by the traditional methods, usually by UV radiation. This work is of great importance because it showed that by using doped fibers instead of standard fibers the sensitivity is increased and for sensing purposes it is a relevant issue.

ACKNOWLEDGEMENTS

M.G. Pulido-Navarro wishes to thank CONACYT for the studentship towards her Ph.D. studies. J.A. Alvarez-

Chavez would like to thank CONACYT, CIITEC-IPN and CONACYT Ciencia Basica, for the funding of his projects.

REFERENCES

1. Lemarquand, V. (1999) Synthesis study of Magnetic Torque Sensors. IEEE Transactions on Magnetics, 35, 4503-510.

2. Wang, Y. and Rao, Y. (2004) Long Period Fibre Grating Torsion Sensor Measuring Twist Rate and Determining Twist Direction Simultaneously. Electronics Letters, 40, 3.

3. Pinet, E., Hamel, C., Glisic, B., Inaudib, D. and Mironic, N. (2007) Health Monitoring with optical Fiber Sensors: From Human Body to Civil Structures. SSN10 Conference Health Monitoring of Structural and Biological Systems, San Diego, 18-22.

4. Hotate, K. (2012) Distributed Fiber Sensing Technology: Currents and Challenges. Optica Pura y Aplicada, 45, 63-69.

5. Peled, Y., Motil, A. and Kressel, I. (2013) Monitoring the Propagation of Mechanical Waves Using an Optical Fiber Distributed and Dynamic Strain Sensor Based on BOTDA. Optics Express, 21, 10697-10705.

6. Heiman, D., Hamilton, D. and Hellwarth, R. (1979) Brillouin Scattering Measurements on Optical Glasses. Physical Review, 19, 6583-6592.

7. Inaudi, D. and Glisic, B. (2006) Distributed Fiber Optic Strain and Temperature Sensing for Structural Health Monitoring. The Third Int'l Conference on Bridge Maintenance, Safety and Management, Portugal, 8.

8. Ladicicco, A., Paladino, D., Pilla, P., Campopiano, S., Cutolo, A. and Cusano, A. (2012) Long Period Gratings in New Generation Optical Fibers. Recent Progress in Optical Fiber Research. Dr. Moh. Yasin (Ed.), Italy, 291-326.

9. Sun, L., Li, J., Jin, L. and Guan, B. (2012) Structural Microfiber Long-Period Gratings. Optics Express, 20, 18079-18084.

10. Bhatia, V. (1999) Applications of Long-Period Gratings to Single and Multi-Parameter Sensing. Optics Express, 4, 457-466.

11. Rao, Y. (2012) OFS Research over the Last 10 Years at CQU & UESTC. Photonic Sensors, 2, 97-117.

12. Zhang, B.W. and Kahiziri, M. (2007) High-Temperature Resistance Fiber Bragg Grating Temperature Sensor Fabrication. Sensors Journal, 7, 586-591.

13. Wang, Y., Wang, D. and Jin, W. (2006) CO2 Laser-Grooved Long Period Fiber Grating Temperature Sensor System Based on Intensity Modulation. Applied Optics, 45, 7966-7970.

14. Vengsarkar, A., Lemaire, P., Judkins, J., Bhatia, V., Erdogan, T. and Sipe, J. (1996) Long-Period Fiber Gratings as Band-Rejection Filters. Journal of Lightwave Technology, 14, 58-65.

15. James, S. and Tatam, R. (2003) Optical Fibre Long-Period Grating Sensors: Characteristics and Application. Measurement Science and Technology, 14, 49-61.

16. Ceballos-Herrera, D., Torres-Gómez, I., Martínez-Ríos, A. and Sánchez-Mondragón, J. (2009) Higher-Order Core Mode Resonances in a Mechanically Induced Long-Period Holey Fiber Grating. Optical Review, 16, 622-626.

17. MacDougall, T., Pilevar, S., Haggans, C. and Jackson, M. (1998) Generalized Expression for the Growth of Long Period Gratings. IEEE Photonics Technology Letters, 10, 1449-1451.

18. Zhu, T., Rao, Y., Wang, J. and Song, Y. (2007) Strain Sensor without Temperature Compensation Based on LPFG with Strongly Rotary Refractive Index Modulation. Electronics Letters, 43, 1132-1133.

19. Kashyap, R. (1999) Fibre Bragg Gratings. Academic, New York.

CHAPTER 10

A Novel Optical Fiber Sensor for Steel Corrosion in Concrete Structures

Christopher K.Y. Leung, Kai Tai Wan * and Liquan Chen

Department of Civil Engineering, the Hong Kong University of Science and Technology, Hong Kong SAR.

ABSTRACT

Steel corrosion resulting from the penetration of chloride ions or carbon dioxide is a major cause of degradation for reinforced concrete structures,. The objective of the present investigation was to develop a low-cost sensor for steel corrosion, which is based on a very simple physical principle. The flat end of a cut optical fiber is coated with an iron thin film using the ion sputtering technique. Light is then sent into a fiber embedded in concrete and the reflected signal is monitored. Initially, most of the light is reflected by the iron layer. When corrosion occurs to remove the iron layer, a significant portion of the light power will leave the fiber at its exposed end, and the reflected power is greatly reduced. Monitoring of the reflected signal is hence an effective way to assess if the concrete environment at the location of the fiber tip may induce steel corrosion or not. In this paper, first the principle of the corrosion sensor and its fabrication are described. The sensing principle is then verified by experimental results. Sensor packaging for practical installation will be presented and the performance of the packaged sensors is assessed by additional experiments.

KEYWORDS

Steel corrosion; iron thin film; ion sputtering technique; optical fiber

1. INTRODUCTION

Reinforced concrete structures make up a large portion of the physical infrastructure of the world, and their durability is an issue of great concern. Steel corrosion is a major culprit for the degradation of concrete members. In order to perform proper maintenance which can extend the lifetime of concrete structures, techniques for steel corrosion detection need to be developed. Within the alkaline environment of concrete, steel is in a passivated state, with a negligible corrosion rate, but when carbon dioxide and chlorides penetrate through the concrete cover to reach the steel, depassivation occurs and the corrosion rate becomes significant. One approach to assess the corrosion of steel is hence to measure the depth of carbonation or the chloride profile in the concrete structure. The conventional approach is to perform chemical analysis on samples cored from the concrete structure. A detailed description can be found in [1]. For carbonation, phenol-phthalate solution is often applied to the sample to identify the depth where color change occurs. However, since the pH at which color change occurs is not the same as the pH below which steel corrosion may occur, some subjective judgment is required for interpreting the results. For chloride penetration, the cored sample is further sliced into thin sections, and each section is ground into powder for chemical analysis. While this can provide an accurate chloride profile in the structure, a major limitation should be pointed out. The chloride concentration does not have a direct correlation with steel corrosion [2]. The critical chloride concentration beyond which steel is depassivated depends also on the pH of the concrete, and this exact relation is not known. Indeed, different researchers [3-4] have come up with different relations between the critical chloride concentration and pH. As another major drawback, taking cores from a large concrete structure at critical locations is often difficult and costly. As a result, the coring operation can only be occasionally performed.

Embeddable corrosion sensors have been developed and installed in concrete structures. The most notable example is the 'ladder' sensor developed in Germany [5]. Parallel steel rods are fixed across two parallel rails to form a small ladder. The ladder is then placed inside a concrete structure in such a way that the rods are at different depths from the surface. By comparing the potential of each rod with a reference electrode, one can assess if corrosion has occurred at the level of a particular rod. The penetration of the corrosion front is then monitored. Note that a major advantage of this technique is the actual assessment of corrosion activity, rather than measuring a variable (such as chloride concentration) for indirect inference of whether corrosion has occurred. A disadvantage of the technique is the relatively high cost associated with the sensing system, and the fact that the 'ladder' sensor can only be installed in new structures. Recently, a cylindrical sensor, based on the same principle has been developed [6]. The cost is still relatively high. Also, when a hole is cored in an existing concrete structure for installing the cylinder, the prevention of preferential chloride penetration along the sensor/concrete interface is an important issue that needs to be further studied.

Optical fiber sensors for chloride detection have also been developed [7]. A particular chloride sensitive solution or indicator paper is brought into contact with concrete. Broad-band light is then sent from one optical fiber through the solution (or paper) to another fiber. In the presence of chlorides, formation of precipitates in the solution or color change of the paper leads to changes in the intensity or spectrum of the transmitted light. For this sensor, an ingenious installation technique, which allows the sensor to be changed over the life span of the structure, has been developed. A limitation of the sensor is that it cannot be applied to existing concrete structures. Also, it is quite large in size, so if a large number of sensors are to be employed to monitor chloride profile at various locations, the installation becomes very inconvenient. Another fiber optic sensor was developed in [8]. The cladding of the optical fiber is made with a special polymer that is sensitive to the environment. Changes in chloride concentration will lead to a change in the properties of the fiber cladding. The corresponding change in the evanescent wave can be monitored through optical spectroscopy. While this is certainly a powerful technique, the principle is rather complicated, and the results may need to be interpreted by a specialist. Also, both optical techniques suffer the limitation that the measurand is the chloride concentration, which does not have a one-to-one relation with corrosion activity [2].

If there is penetration of carbon dioxide or other acids into concrete, steel corrosion starts when the pH drops below 11.5. In [9], a fiber optic pH sensor is developed to monitor pH values between 9 and 12, which covers the critical pH value below which corrosion will occur. The sensor is based on the absorption intensity ratio between two different wavelengths at a pH sensitive membrane. In [10], a different fiber optic pH sensor is reported. An optical fiber is placed inside a polymer matrix with covalently coupled dye. When the pH values vary between 11 and 12.5, the color change of the dye is monitored with a micro-fabricated electro-optical sensing system. Using a different type of dye, a humidity sensor based on the same principle has also been developed [10]. While the measurement of pH and humidity is useful in identifying corrosion caused by loss of alkalinity in the concrete environment, chloride-induced corrosion cannot be revealed.

In this paper, we will present a new fiber optic based technique for detecting the onset of steel corrosion under a certain environment, with special application on concrete structures. The physical principle is first explained. The sensor will be tested in sodium chloride solution and embedded in cement mortar with sodium chloride to illustrate the feasibility of corrosion sensing. Then, sensor packaging for practical application will be presented. Additional tests were conducted on the packaged sensor to study its performance. First, sensors are installed inside holes drilled in hardened cement mortar blocks with different sodium chloride content to verify the capability to detect corrosion activity in existing structures. Second, sensors are glued to embedded steel reinforcements under corrosive environment and the sensor output will be compared to corrosion current measurement. Third, packaged sensors are embedded in cement mortar blocks with the sensing tip at different distances from the surface. With these blocks placed inside a salt water bath, the feasibility to monitor the penetration of chloride into a concrete member, in a way similar to the corrosion "ladder" sensor [5], can be assessed.

2. PHYSICAL PRINCIPLE

The novel corrosion detection method is based on the simple principle of light reflection illustrated in Figure 1. An optical fiber is stripped and cut with a cleaver to produce a flat surface at its end that is perpendicular to the fiber axis. Using the ion sputtering technique, an iron thin film (thickness around one hundred to several hundred nanometers) is deposited on the cleaved fiber end. Light is then sent through the optical fiber and the reflected light intensity is monitored as a function of time. Initially, the iron film at the end of the fiber acts like a mirror and all or most of the light is reflected (Figure 1a). As corrosion occurs, the film thickness decreases with time. Consequently, the reflected signal also drops (Figure 1b). Ultimately, after the film is completely removed, the reflected light intensity will return to the level for the cleaved glass surface.

Compared to other corrosion sensors, the 'reflection' sensor possesses a number of advantages. Since the principle is very simple, interpretation of results is easy and direct. It is applicable to corrosion induced by either chloride penetration or carbonation. The technique does not require the use of special optical fibers, so low-cost single mode telecommunication fibers can be employed. In the sputtering process, a large number of fibers can be placed in the sputtering chamber and coated at the same time. Mass production of the sensor at low cost should therefore be possible. Moreover, with the small size of the sensor, it can be retrofitted to existing structures. Two plausible ways to install the sensor are illustrated in Figure 2.

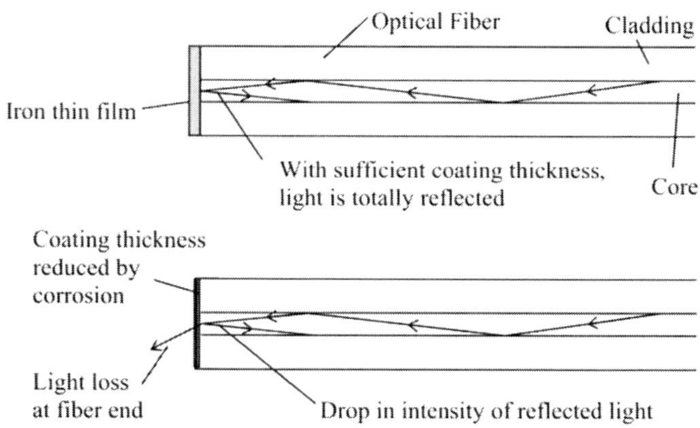

Figure 1. Illustration of the corrosion sensing principle.

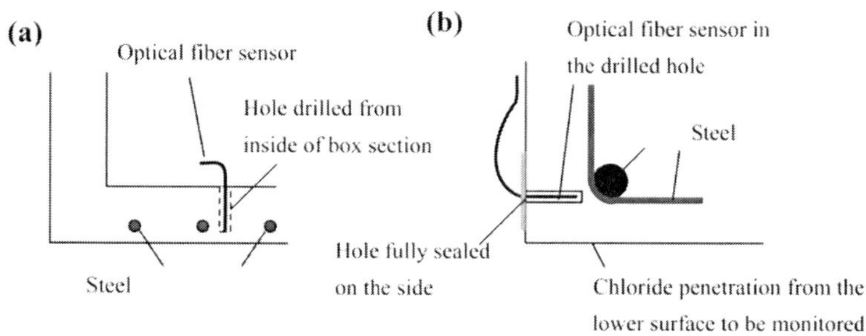

Figure 2. Plausible ways for retrofitting the sensor on an existing structure with (a) boxed section, (b) solid section.

For a bridge with boxed section, a hole of required depth can be drilled from the inside of the box for placement of the sensor (Figure 2a). To monitor chloride or carbon dioxide penetration from a surface of a solid section, the hole can be drilled on a perpendicular surface and sealed (Figure 2b). With these approaches, the drilled hole will not act as an easy diffusion path that will affect the monitoring results. Since the monitoring involves light intensity measurement, to avoid variations due to fluctuations in the light source, and changes in connector efficiency when fibers are disconnected and re-connected, the loss in reflected light power is best measured with an optical time domain reflectometer (OTDR). This method measures the relative power of light reflected at points along the fiber, and that at the fiber end. Since it is a relative measurement, results will not be affected by the variations mentioned above. From our experience, single mode optical fiber is superior to multimode fiber in terms of measurement stability. This is because the intensity of reflected light at the film depends on the incident angle of the ray, which varies for different modes. For a multimode fiber, any small deformation along the fiber may lead to mode redistribution that causes fluctuation of the sensor response. Our tests are conducted in 850nm wavelength, as this is the operating wavelength of our OTDR. The cost of 850nm single mode optical fiber and its accessories is much higher than those of 1550nm single mode optical fiber (which is the common type for communication applications). As the operating wavelength is not an essential parameter of the sensor, an OTDR system and single mode optical fiber at 1550nm should be employed in practical applications. However, as light absorption by the film is wavelength dependent, the sensor should be calibrated at the wavelength at its operation. Also, the spatial resolution of the sensor does not affect the sensor performance as long as the fiber end is sufficiently far away from the bulkhead of the OTDR. Hence, a portable OTDR with low cost (around US$6000) can be used for the measurement.

3. SENSOR FABRICATION AND VERIFICATION OF THE DETECTION PRINCIPLE

To form an iron thin film with controlled thickness on the cleaved end of the optical fiber, the sputtering deposition technique was employed. The fibers were placed inside a vacuum chamber, with the cleaved ends facing a pure iron target. The sputtering process involves the bombardment of target with energetic ions to induce the ejection of vapor particles. Inside the vacuum, the vapor will deposit as a thin film on the surface of the fiber. The film thickness depends on the distance of the fiber from the target as well as the deposition time. Before the coating of fibers, a calibration experiment was performed using a glass plate with a strip of thin tape bonded to its surface. The plate was placed at a distance to the target that was equal to the fiber/target distance in the actual sputtering process. Sputtering deposition was performed for different periods of time. After the deposition process, the tape was removed and the thickness of coated film was obtained from the surface profile along the coated plate, measured with a profileometer. With the calibration experiment, the relation between film thickness and deposition time was found to be linear and the slope was 2.92nm/min.

Based on the aforementioned sensing principle, a bundle of cleaved fiber (Nufern 780-HP) was put into the sputtering chamber facing the pure iron target. With different deposition times, fibers with different coating thicknesses were prepared. In this study, the definition of reflectivity is 10log(Proportion of light reflected). The reflectivity of cleaved fiber end without any coating is calculated theoretically to be -14.6dB (the refractive index of the core of the optical fiber is 1.458). For fibers with various thicknesses of iron coated at its tips, the reflectivity was experimentally determined by an OTDR (Opto-Electronics, OFM-20). The results (Figure 3) indicate significant difference (about 8dB) in reflectivity between the coated and uncoated fiber. Also, the reflectivity seems to approach an asymptotic value (about -6dB) for film thickness greater than 300nm. There was no reflectivity data for film thickness in the range of 25nm to 100nm. The dotted lines represent a hypothetical bilinear relationship. There is a sharp change in the sensitivity of the sensor at about 60nm film thickness.

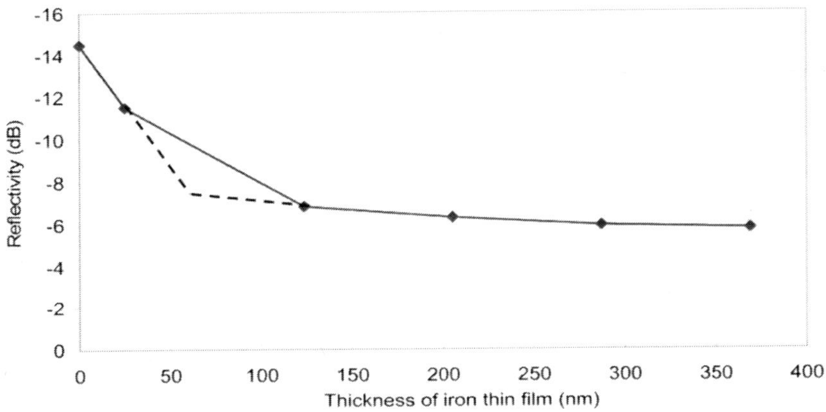

Figure 3. Relationship between the reflectivity at the coated fiber end and the film thickness.

The first verification experiment was performed using 100 nm iron film on their tips. The coated end of optical fibers were dipped into calcium hydroxide solutions (pH=13 to simulate the concrete environment) with 0.1, 0.5 and 1.0 mol/L of sodium chloride, respectively. The reflected signal from the fiber end was monitored daily with OTDR. The variations of reflectivity (in dB) with time (in days) under different chloride concentrations are shown in Figure 4. It should be pointed out that an optical fiber sensor has also been placed in alkaline solution without any sodium chloride added, and no change in optical power was observed over two months. Based on the test results, two observations can be made. First, the presence of a corrosive agent (chloride ions) leads to a drop in reflected optical power with time, which can be attributed to the gradual removal of the iron film on the fiber end. Second, the power drop which represents the total corrosion of the film increases with increasing chloride concentration. These findings demonstrate the feasibility of the proposed sensing concept.

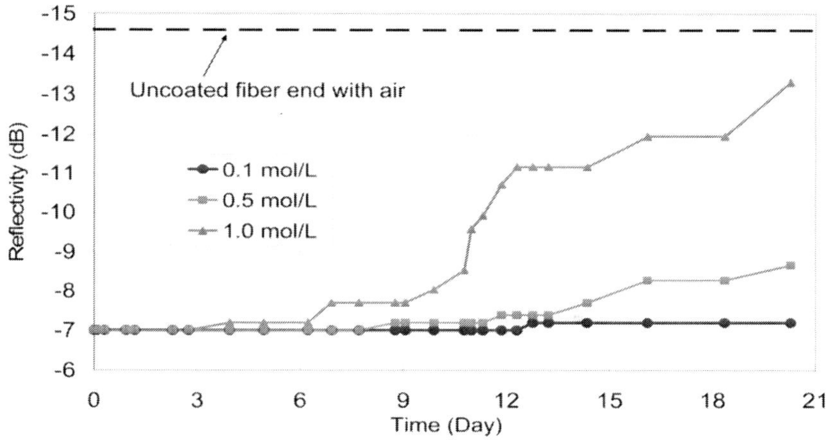

Figure 4. Effect of chloride concentration on corrosion sensor output.

From the test results, the measured loss does not increase smoothly with time, but may show a sudden jump followed by a period of little or no increase. This is most obvious for the case with 1.0 mol/L of sodium chloride. Actually, what is monitored is the film thickness at the core of the optical fiber with a small area of only 64 μm^2. In other words, we are measuring the corrosion activity within a small zone, rather than the overall corrosion over a reasonably large region (such as in macrocell current measurements). The results are therefore very sensitive to the local chloride concentration and moisture content, which can vary with time. In particular, once reaction occurs, chlorides are used up and diffusion of additional chlorides to the region takes time. This explains the large reduction in corrosion rate (indicated by constant signal loss versus time) after a rapid increase in corrosion activities.

A second set of experiments was performed to investigate the corrosion sensing capability inside an actual concrete environment. The coated optical fibers were cast inside small blocks of cement mortar with water:cement:sand ratio of 0.5:1:2. Sodium chloride was added at 5% of the cement weight. After curing for 7 days, two specimens were placed inside a warm water bath at 30°C. Another two specimens were taken out of the curing room and exposed to the dry laboratory environment. For each specimen, the reflectivity of the end of the embedded optical fiber sensor was monitored as a function of time. The results are shown in Figure 5 for both the wet and dry specimens. When kept wet, corrosion (which is indicated by the drop of reflectivity) started much earlier and occurs to a much larger extent. For the specimen in the laboratory environment, which is relatively dry but not completely dry, corrosion started to occur after two months, and did not reach a high level. The findings indicate that the sensor can detect chloride ions inside the concrete environment, and it can distinguish between the seriousness of corrosion activities under different environments.

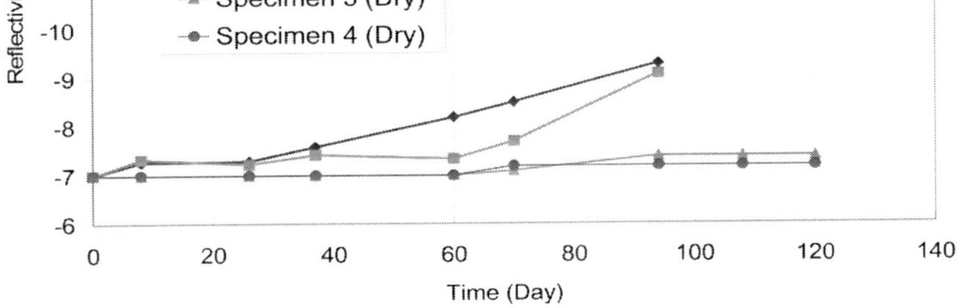

Figure 5. Output from sensor embedded in concrete under different environments.

4. PACKAGING OF THE SENSOR

The sensor employed in the above experiments is simply a bare fiber with an iron coating. The installation of sensor into concrete has to be carried out with great caution to avoid fiber breakage. While this is achievable in the laboratory, field application of the sensor requires proper packaging to make it robust against the concrete casting process. An approach to package the sensor is illustrated in Figure 6. To fit the fiber bundle into the sputtering chamber, the length of each coated optical fiber is about 200 mm. The coated fiber is embedded in a 1 mm diameter hollow flexible plastic tube and fixed inside by glue. The coated end inside the plastic tube is then well protected (from mechanical actions) during the concrete casting process.

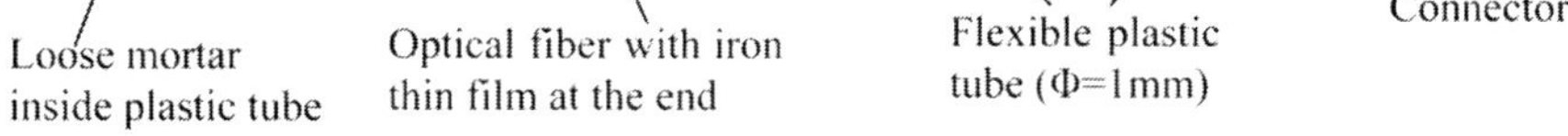

Figure 6. Schematic diagram of sensor packaging.

To enable proper transfer of corrosive agents from the concrete to the tip of the fiber, mortar is added into the plastic tube (see Figure 6 and inset of Figure 7). A proper mortar mix can be made with fine sand (diameter less than 0.2 mm) and high water/cement ratio (w/c=0.6). A high water/cement ratio is necessary to ensure the easy transportation of corrosive agents to the end of the optical fiber. To get the mortar into the fine plastic tube, a bundle of tubes with optical fiber (about ten per batch) is inserted into mortar with the tube end just under the mortar surface (Figure 7).

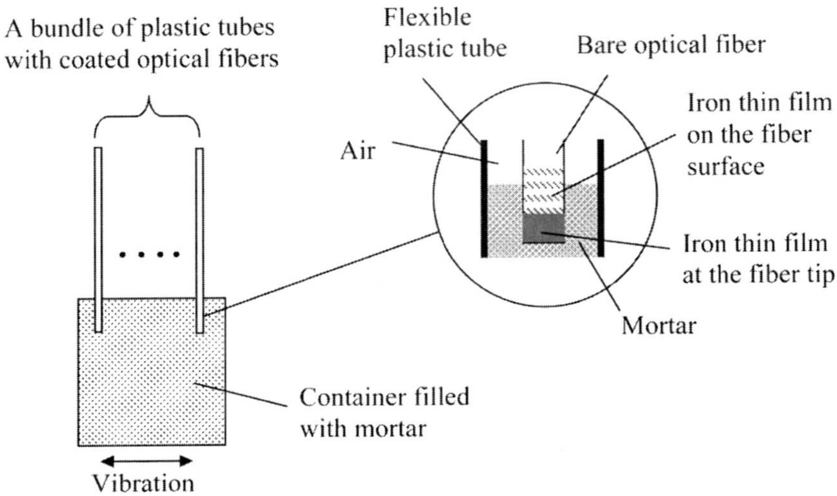

Figure 7. Fiber sensor packaged inside a mortar block.

For proper filling, the mortar container is shaken on a vibration table to facilitate the penetration of mortar into the plastic tube. The filled plastic tube is then taken out and placed horizontally until the mortar is hardened. After mortar filling, the fiber end should be surrounded completely by mortar (see the inset of Figure 7), and this has been verified by cutting some of the plastic tubes to reveal the mortar and fiber inside. While the optical fiber is surrounded by mortar (which is alkaline), it would not be attacked as the glass surface is covered with iron thin film during the sputtering deposition process. The length of the coated optical fiber, which is only 200 mm, is certainly not long enough in practical applications. It is therefore connected to an extension cable by fusion splicing and the joint is protected by a heat-shrink splice protector (Figure 6). The whole fiber (coated as well as fiber extension) is protected by flexible plastic tube and no bare fiber is exposed. For use in harsh environment of the construction site, the extended fiber can be further protected by an armored loose tube optical fiber cable.

5. EXPERIMENTS ON PACKAGED SENSORS

5.1 Experimental set-up and procedure

In Section 3, the sensors were tested on their own to show that they can respond to chlorides in the environment. In real applications, however, we need to ensure that the packaged sensor is effective in providing information on steel corrosion inside a concrete structure. In the following, the experimental set-ups and procedure for three separate tests on the packaged sensor in Section 4 will be described in detail.

5.1.1 Sensors in hardened cement blocks

The possibility to post-install the proposed sensor in existing structures is one of its major advantages over other available sensors. To verify such capability of the sensor, cement mortar cubes (40 mm × 40 mm × 40 mm) were first cast with 0.4%, 1% and 2% of sodium chloride by the cement weight (see Figure 8a).

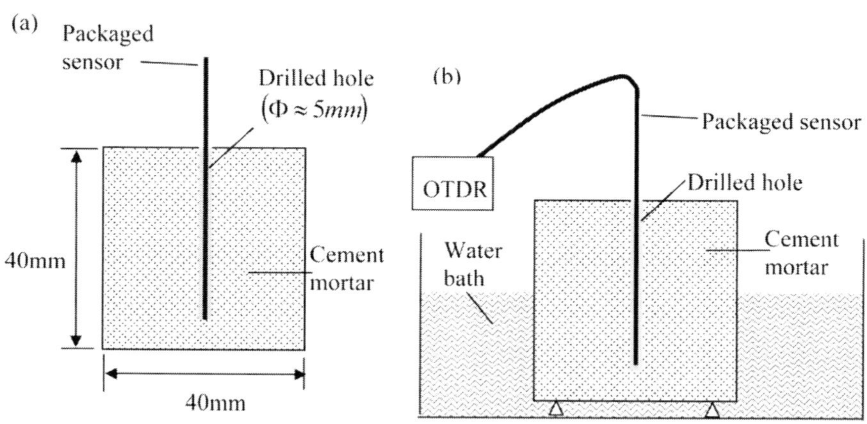

Figure 8. Experimental set-up of the test of sensors in drilled hole of a hardened cement mortar block.

The mix proportion of the cement mortar was water:cement:sand ratio=0.6:1:2. After curing the specimens for 14 days, a 5 mm diameter hole was drilled into each specimen to reach a depth that was 5 mm from the bottom. The hole was cleaned by water and a packaged sensor was inserted into the hole. In the tests, the coated film on the fiber tip was about 200 nm in thickness, and the tube diameter was about 1 mm. The hole was then filled by cement mortar with the same mix proportional as the specimen (but without sodium chloride). In practice, the cement mortar for filling the hole should be at least as porous as the parent material, so the water/cement ratio should be the same or higher. Also, a hole size of 5 mm is appropriate, as a smaller hole will be difficult to be filled. After the sensor was installed and the hole was filled, the specimens were put into a water bath to keep the moisture inside the specimens (Figure 8b). As shown in the figure, the bottom of the specimen and part of the vertical surfaces were exposed to water during the test.

5.1.2 Modified ASTM G109 test

The objective of this experiment is to compare the sensor response to a different approach of corrosion detection, to see if consistent results can be obtained. The chosen approach of comparison is the measurement of corrosion current, which is a standard method described in ASTM G109 [11]. The employed experimental set-up is shown in Figure 9 with all the relevant dimensions.

A salt water reservoir with 3% sodium chloride solution was placed on top of a concrete prism, to provide a source for chloride ions that would penetrate into the concrete specimen. A steel reinforcement, 10 mm in diameter, was placed near the top surface as an anode with a cover of 10 mm. Two additional steel rods were also placed close to the bottom of the member as cathodes with a cover of 25 mm. All the steel reinforcement was covered by tape, except for the portion under the salt water reservoir. All the vertical surfaces of the concrete prism were sealed by epoxy to ensure a one-dimensional diffusion path. At the top of the steel reinforcement facing the upper surface, a packaged optical corrosion sensor was attached. The thickness of iron film on the fiber was about 200 nm. The specimens were cured for 14 days before the salt water reservoir was installed. During the test, the reflectivity of the sensor was measured at different times. Also, the potential difference over a 100 Ω resistor (AB in Figure 9a) between the anode and cathode was measured to calculate the macrocell current. Since the bars at the bottom of the specimen should not corrode, an increase in macrocell current indicated the onset of corrosion at the upper bar. The threshold of corrosion macrocell current is about 10 μA [1]. The corrosion initiation times obtained from the drop of reflectivity of the sensor and the rise of macrocell current can be compared to see if the two methods give rise of consistent results.

It should be pointed out that our experimental set-up was identical to the one recommended by ASTM G109 except that the cover of the anode rebar was reduced from 25 mm to 10 mm, so that the corrosion process could be accelerated. As both diffusion and absorption occurs in the first 10-20 mm of concrete from the surface during wet/dry cycle [12,13], a test with small concrete cover is not suitable for the comparison of material parameters from one mix to another (as the transport process is governed by more than one parameter). However, since the objective of this experiment is simply to compare the corrosion initiation time indicated by the sensor and the macrocell current, the use of non-standard set-up is acceptable.

5.1.3 Diffusion test

For long term monitoring, it is necessary to know the penetration front of corrosion activity. The present experiment aims at verifying the potential of the proposed corrosion sensor to monitor the penetration of corrosive agents, such as chloride ions, carbon dioxide and sulphur dioxide. In this test, three packaged sensors were embedded into a cement mortar cube (40mm× 40mm× 40mm) with the tip at different distances (5mm, 10mm and 15mm) to the surface (Figure 10). In this experiment, the thickness of the coated iron thin film on the fiber was about 400nm. During the casting of mortar to make the blocks, toothpicks were used to hold the fiber in place. After the blocks were cast and cured, epoxy was applied on all vertical sides so the corrosive agents and water can only penetrate from the bottom. The specimen was then put into a salt water bath with 3% sodium chloride. The reflectivity of the three fibers was then measured as a function of time. Note that the concentration may change due to evaporation and diffusion. Although the corrosion initiation time may not be sensitive to the chloride concentration of the source as long as it is larger than 1% [14], the salt water bath was replaced by a new solution every two weeks to maintain a constant solution during the test.

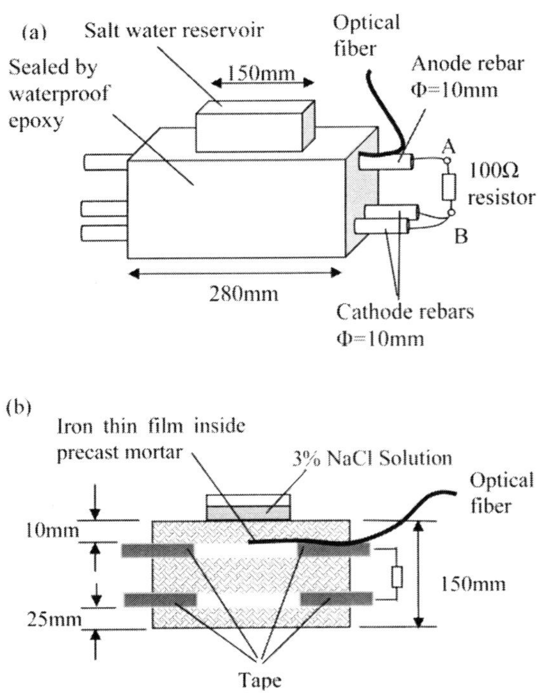

Figure 9. Set-up to compare optical sensor output with corrosion current of steel.

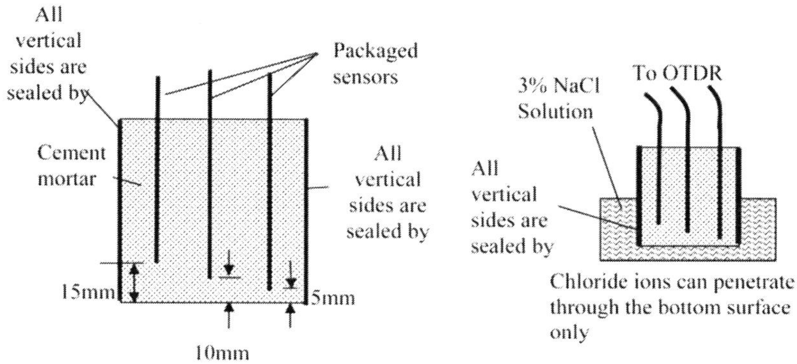

Figure 10. Set-up of diffusion test.

In practice, we are interested to knowing if corrosion has started in the reinforcing steel inside a concrete structure. Several optical sensors can be placed inside the concrete member with their tips at a smaller distance from the surface than the steel bar. Monitoring of corrosion activities indicated by the sensors can provide sufficient warning to the corrosion of steel reinforcement, and proper maintenance action can be taken.

5.2 Experimental results and discussion

In this section, the experimental results of the tests in Section 5.1 are presented to demonstrate the viability of the proposed optical corrosion sensor. The output from the optical sensor is converted to a reflectivity value. When the coated iron film has not been completely removed, the fiber end is isolated from the environment and the reflectivity versus film thickness relationship should follow the calibration curve in Figure 3. Once the whole thin film is removed, the reflectivity depends on the surrounding environment, which can give rise to two different situations. In the first situation, the fiber end is in contact with air, and the reflectivity is -14.6 dB. In the second situation, the fiber end is surrounded by water (when the pores inside concrete are saturated), and the theoretical reflectivity reaches a much higher value of -26.8 dB.

5.2.1 Sensors in hardened cement mortar blocks

Packaged sensors were inserted into holes in cement mortar blocks with different concentrations of sodium chloride. The results are shown in Figure 11. The reflectivity of the sensor embedded into 1% and 2% sodium chloride was reduced after about 40 days of embedment, while the reflectivity remained at the same level for the 0.4% case. The threshold of chloride concentration for steel corrosion is about 0.4% for structural steel [1]. For the sensor, corrosion of pure iron at its end was measured. At 0.4% sodium chloride, there was nearly no change in reflectivity in a period of three months. For the other two cases (with 1% and 2% sodium chloride, which are both above the threshold), significant drop in reflectivity was found to occur at about the same time (after 46 days). In these tests, the sensors were not directly cast into the mortar blocks, but inserted into a drilled hole which was later filled with mortar (without chloride). The results indicate that corrosion activities in a concrete member with corrosive agents can be detected by a post-installed sensor. The potential for corrosion monitoring in existing structures is hence demonstrated.

If one takes a reflectivity of -14.6 dB to indicate the complete corrosion of the iron film on the fiber, the corrosion rate appears to be higher for the case of 1% sodium chloride. The results indicate that corrosion initiation and corrosive rate may not be sensitive to the chloride concentration, as long as it is sufficiently high. After the reflectivity goes below -14.6 dB, large fluctuations can be observed for the 2% sodium chloride case. A plausible explanation is the presence of cement hydration products right around the fiber tip that affects the reflectivity. This phenomenon may be similar to the effect of soil particles on in-situ spectroscopic measurements reported in the literatures [15, 16].

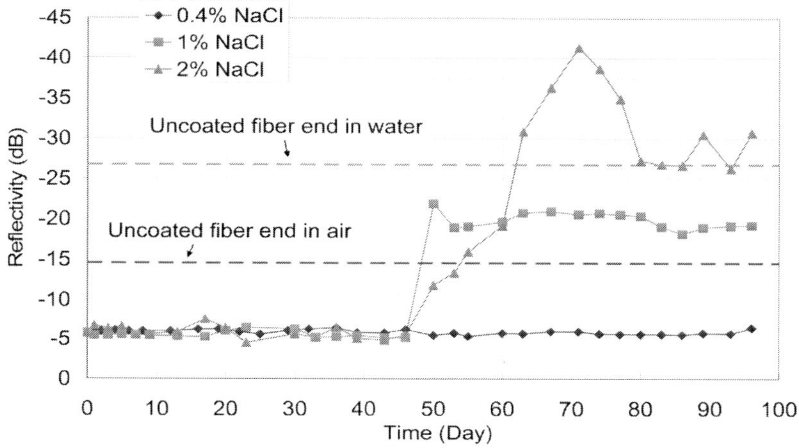

Figure 11. Results of tests of drilled hole of hardened cement mortar block.

5.2.2 Modified ASTM G109 Test

By the tests of drilled hole of hardened cement mortar block, the packaged sensors are shown to be responsive to chloride ions. Here, the corrosion initiation time obtained from the sensors is compared to the macrocell current measurement by the standard ASTM G109 test. The results are shown in Figure 12. For the first specimen (Figure 12a), the macrocell current indicated the start of steel corrosion (when the macrocell current reaches 10 μA) at about the 50th day while the decrease of reflectivity was at about the 55th day. The difference is about 5 days. The time of corrosion initiation from the optical sensor and the macrocell current was very close. For the second specimen (Figure 12b), the reflectivity of the optical sensor showed significant drop at about the 42nd day while the macrocell current indicated that the steel corrosion started at about the 59th day. The difference is about 17 days. It should be pointed out that the difference in initiation time, which is significant for the experiment, may not be significant in practical applications. In practice, the initiation time of steel corrosion is over several years. After initiation, corrosion progresses at very slow rate. Even if the corrosion is detected a few months after it starts, there is little consequence.

(a)

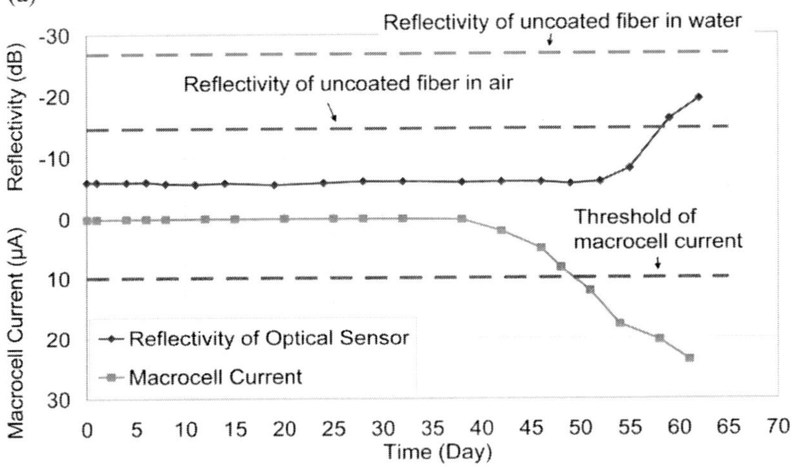

(b)

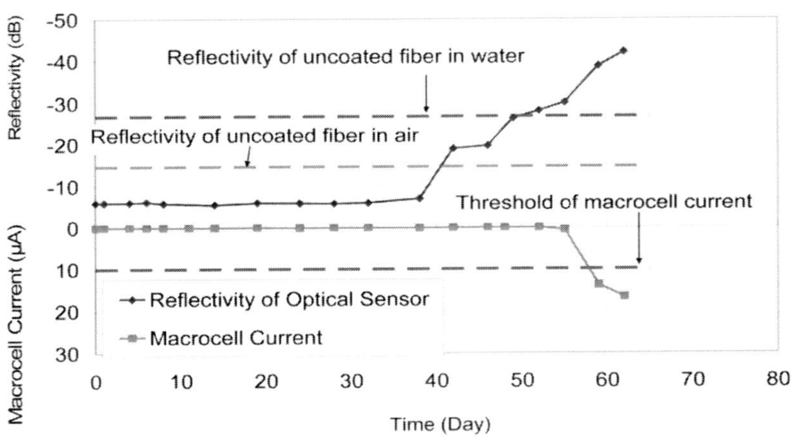

Figure 12. Results of standard ASTM G109 tests. (a) 1st specimen, (b) 2nd specimen.

5.2.3 Diffusion test

In this test, the penetration front of chloride ion is monitored by measuring the output of sensors at different distances to the specimen surface. The results are shown in Figure 13. For the sensor at 5 mm from the surface, the reflectivity started to drop at the 47th day, while the iron film was completely removed at the 77th day (when the reflectivity dropped to -14.6dB). The sensor at 10mm from the surface showed a significant drop of reflectivity at about the 67th day, while the reflectivity of the sensor at 15 mm from the surface remained constant up to about 90 days. It should be noted that the coated film thickness in this experiment is 400nm while that in the previous experiment is around 200 nm. Therefore, it is not surprising that the corrosion initiation time in this test is higher than that in the previous one, even through the distance of the fiber from the surface is smaller.

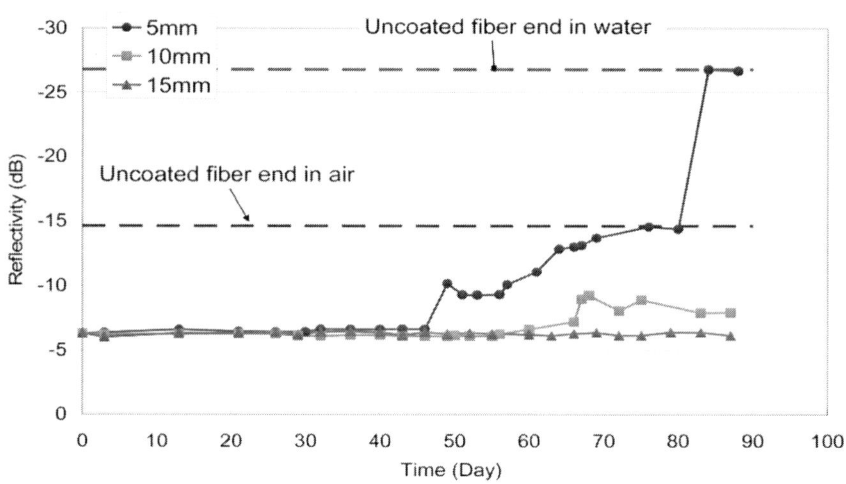

Figure 13. Results of diffusion tests.

The test results, which show earlier initiation of corrosion at a smaller distance from the source of chlorides indicates the feasibility of corrosion front monitoring with the packaged sensors. However, the current results are mostly qualitative in nature. Further investigations are required to relate the corrosion initiation time to the distance from the surface, so theoretical extrapolation can be performed with measured results to predict the future penetration rate of the corrosion front. Also, experiments should be performed to study the response of the sensor under other corrosive environments, such as concrete with penetrated carbon dioxide or sulphur dioxide.

6. CONCLUSIONS

In this paper, we have presented a new fiber optic sensor for the detection of steel corrosion in concrete structures. The physical principle is based on the corrosion of an iron thin film at the end of an optical fiber, which will reduce the reflectivity from the fiber end. The sensing concept is verified by experiments. Practical sensor packaging method is proposed for embedding the sensor into existing and new concrete structures. Experimental results indicate that the sensor can detect the presence of chloride ions and the change in sensor output is dependent on the chloride concentration or moisture condition of the surrounding environment. The corrosion initiation time determine from the drop in sensor reflectivity is consistent with that obtained from standard macrocell current measurement. By placing sensors at different distances from a concrete surface exposed to chloride ions, the potential to monitor the penetration of chloride front inside a concrete member is demonstrated. While further work is certainly necessary to fully understand the correlation between the change in reflectivity and the corrosion process, the available experimental results do illustrate the feasibility and potential of the proposed sensing technique for practical applications.

ACKNOWLEDGMENTS

Financial support of the work by the Hong Kong Research Grant Council, under CERG Project 616405, is gratefully acknowledged.

REFERENCES

1. Broomfield, J.P. Corrosion of Steel in Concrete; E & FN SPON: London, 1997.

2. Bentur, A.; Diamond, S.; Berke, N.S. Steel Corrosion in Concrete; E & FN SPON: London, 1997.

3. Hausmann, D.A. Steel Corrosion in Concrete. Materials Protection **1967**, 6(11), 19–22.

4. Diamond, S. Chloride Concentrations in Concrete Pore Solutions Resulting from Calcium and Sodium Chloride Admixtures. Cement and Concrete Aggregates **1986**, 8(2), 97–102.

5. Schiessl, P.; Raupach, M. Monitoring System for the Corrosion Risk for Steel in Concrete. Concr. Int. **1992**, 14(7), 52–55.

6. Raupach, M. Corrosion Behaviour of the Reinforcement under On-Site-Conditions. In Proceedings of 15th International Corrosion Congress; Frontiers in Corrosion Science and Technology: Granada, 2002; Art.096.

7. Huston, D.R.; Fuhr, P.L. Distributed and Chemical Fiber Optic Sensing and Installation in Bridges. Fiber Optic Sensors for Construction Materials and Bridges **1998**, 79–88.

8. Ghandehari, M. Ingress Monitoring in Concrete Structures. In Proceedings of the 15th ASCE Conference on Engineering Mechanics; New York, 2001; CD-ROM.

9. Habel, W.R.; Hofmann, D. Fibre Optic Sensors for long-term SHM in Civil Engineering and geotechnique. *ISHMII-3 Conference*, Vancouver Canada; Nov. 2007; Art. 191.

10. Melhorn, K.; Flachsbarth, J.; Kowalsky, W.; Johannes, H.H. Novel Sensors for long-term monitoring of pH and humidity in concrete. In Proceeding of the 6th IWSHM Conference; Stanford CA, 2007; pp. 387–394.

11. ASTM Standard G109. Standard Test Method for Determining Effects of Chemical Admixtures on Corrosion of Embedded Steel Reinforcement in Concrete Exposed to Chloride Environments. ASTM International; West Conshohocken, PA, 2007. www.astm.org.

12. Weyers, R.E.; Prowell, B.D.; Sprinkel, M.M. Concrete Bridge Protection, Repair, and Rehabilitation Relative to Reinforcement Corrosion: A Methods Application Manual **1993**. SHRP-S-360.

13. Tuutti, K. Chloride induced corrosion in marine concrete structures. In Durability of Concrete on Saline Environment; Uppsala, 1996; pp. 81–93.

14. Zhang, J.Y.; Lounis, Z. Sensitivity analysis of simplified diffusion-based corrosion initiation model of concrete structures exposed to chlorides. Cem. Concr. Res. **2006**, 36, 1312–1323.

15. Sinfield, J.V.; Germaine, J.T.; Hemond, H.F. Effects of soils on laser induced fluorescence of BTX contaminated pore waters. In Journal of Geotechnical and Geoenvironmental Engineering; ASCE, 1999; Volume 125, 12, pp. 1072–1077.

16. Bogrekci, I.; Lee, W.S. Design of a portable Raman sensor for phosphorus sensing in soils. In ASAE Meeting Presentation, ASAE Annual International Meeting, Tampa, Florida, 17 - 20 July; 2005; p. 051040.

Index